EMPIRE
OF
LIGHT

a history of discovery in science and art

Sidney Perkowitz

JOSEPH HENRY PRESS
WASHINGTON, D.C.

Joseph Henry Press • 2101 Constitution Avenue, N.W. • **Washington, D.C. 20418**
www.nap.edu

The Joseph Henry Press, an imprint of the National Academy Press, was created with the goal of making books on science, technology, and health more widely available to professionals and the public. Joseph Henry was one of the founders of the National Academy of Sciences and a leader of early American science.

Any opinions, findings, conclusions, or recommendations expressed in this volume are those of the author and do not necessarily reflect the views of the National Academy of Sciences or its affiliated institutions.

Library of Congress Cataloging-in-Publication Data
Perkowitz, S.
Empire of light: a history of discovery in science and art/Sidney Perkowitz.
p. cm.
Includes bibliographical references and index.
ISBN 0-309-06556-9
1. Light.   I. Title.
QC355.2.P48   1998
535—dc21   98-43181
                                        CIP

To Sandy and Mike, with love

Contents

Preface

Our consciousness is largely occupied by light. We use it throughout our waking existence, and we see it in our dreams; we respond to its purity, to its appearance in painting and photography, to the thrilling thought that it was present at the Creation, be that interpreted as the Big Bang or "Let there be light." I want to tell a tale about the physical nature of light and how we perceive it, about its importance in astronomy and in our technological world, about its artistic significance.

This book comes out of my involvement as a physicist and as a lover of the visual arts, and began decades ago in New York City. When I was ten or twelve years old, my father took me to the American Museum of Natural History, where I first saw beautiful crystalline gems and minerals. Years later, I became a scientist, using laser light to examine pure crystalline solids. And as a young student, I visited the Museum of Modern Art and came upon René Magritte's *Empire of Light*. Long before I knew much about light, I felt its mysteries as expressed in that work. The painting shows a paradox: an ordinary house and street in lamp-lit nighttime darkness, under a blue sky full of daylight. That strangeness within a commonplace scene reminds me that the light we see daily carries unresolved enigmas.

The riddles of light have inspired scientists since the time of the Greek philosophers. Scientific analysis of light has been instrumental in the theories of relativity and of quantum physics, the understanding of human perception and consciousness, the study of the expanding universe and the Big Bang; inventions that create and modify light define the modern world. The science of light is still unfolding. We have not fully grasped how the brain interprets light or how light can exist as both waves and particles. The puzzle of its duality remains unexplained nearly a century after Albert Einstein proposed the photon, the quantum particle of light. His works, and those of Empedocles and René Descartes and Sir Isaac Newton, Thomas Edison and Santiago Ramón y Cajal and Edwin Hubble and Richard Feynman, among others, are high points here.

Artists also ask questions as they observe the world and the universe. Their answers are given in light that they have shaped to convey beauty and meaning. In *Empire of Light,* as I saw the painting at MOMA (or in another version, *L'Empire des Lumières*, appearing on the cover of this book), Magritte created a metaphor for the mysteries of light. Other artists have shown how we see light and how it behaves in nature; still others deal strikingly with color, or display varieties of artificial light, or use light itself as a medium of expression. My story is illuminated by the art of the early cave painters; of Michelangelo da Caravaggio, who painted light for dramatic effect; of the painter Giovanni Pittoni and the architect Etienne Boullée, who honored Newton's researches in light; of Vincent van Gogh, who passionately formed light with pigments that carry scientific meaning; of Edgar Degas and Sonia Delaunay and Edward Hopper and James Turrell, and more.

My hope is that I weave together science and aesthetics, technology and art, to bring you a coherent sense of light that will transform your vision of the world.

It is a pleasure to thank those who have contributed to the writing of this book. The members of my family did more than give moral support.

My wife, Sandy, a self-confessed unscientific humanist, listened patiently and read attentively. She helped me time after time to put scientific ideas into clear prose. My son, Mike, drew upon his new degree in cognitive science from Brown to teach me about the perception of light and to help me with creative library research.

Elaine Markson, my agent, showed immediate enthusiasm for my idea; even better, she knew when to intervene. Jack Macrae, my editor, steadily reminded me of an essential principle—simplify the writing, not the science; his assistant, Albert LaFarge, helped in numerous ways. Peter Brown, editor of the New York Academy of Sciences magazine *The Sciences*, warmly welcomed my essays about light that appeared within the magazine, which helped turn my thoughts toward this book.

Jennifer Barlament, Monique Hosein, Patricia Rabine, and Michael Shattah tenaciously tracked down obscure references and suggested other ones. Friends and colleagues at Emory University, the Fernbank Observatory, and the National Synchrotron Light Source drew on their specialties to critically read portions of the manuscript: Ben Arnold, Krishan Bajaj, Ron Boothe, Ray DuVarney, Henry Edelhauser, Nick Gmur, Jack Kinkade, Pat Marsteller, Rick Williamon, and Gwyn Williams. Also at Emory, Clark Poling, professor of art history; Max Anderson, Director of the Michael C. Carlos Museum; and Judy Zanotti, of Theater Emory, enhanced my ideas about light in art. My students at Emory, in Reading and Writing Physics, and at the Atlanta College of Art heard or read parts of the manuscript. Mark Pozner read and commented with care. Any errors in this book are my doing, not that of these contributors.

Members of the LitSaloon Writers' Group were always there, especially Gloria Brame, my particular friend and trusted adviser. I will miss Russ Fazio's special enthusiasm, cut short by his untimely passing. And those oldest and dearest friends came through again: Jeffrey and Andrea Shoap, and Don Shure. May the light shine on all of them.

Atlanta, Georgia
April, 1996

1

The Birth and Meaning of Light

How far will chemistry and physics . . . help us to understand the appeal of a painting?

—Hazel Rossotti, *Colour: Why the World Isn't Grey*

When I walk into my laser laboratory, I command every kind of light.
I flip the wall switch, and the room fills with the cool blue-white glow
from fluorescent tubes overhead. Next I turn on a desk lamp. Its hot
incandescent bulb makes a warmer, redder kind of everyday light. I
throw another bank of switches, and one of my big lasers comes up.
The glowing gas in its long tube makes an intense emerald green
beam, whose ultra-pure tint nearly sears my color sense. Another hint
of the exotic comes in the invisible ultraviolet light the laser also
makes, its high energy capable of bringing deep harm to living things.
The last set of switches turns on my infrared laser—and wakes
my sense of wonder. That laser's invisible light does not threaten life,
but it still carries powerful magic. It is intimately connected to the
beginning of our universe, to the Big Bang that occurred billions of
years ago.

This sweep, from the cosmic beginnings to my present laboratory, shows that light extends from the large scale of the universe into the ordinary human world. No other single phenomenon crosses so many human and physical categories. The physical understanding of light has involved the most significant of scientific ideas: theories of waves and quantum particles, of relativity and the Big Bang. Its human impact is equally profound: Light determines our existence, occupies much of our thinking capacity, excites our sense of beauty. Each category is important in itself; considered together through the unifying theme of light, they represent a cross-section of our universe.

As I stand in my laboratory, I can place each kind of light by its physical properties and my physiological response: its intensity; whether it is visible or invisible; its wavelength, which for visible light translates into color; its effects on inert matter and on living stuff, which may be essential and benign, or fatal. My less rational side knows light at a deeper level not so amenable to analysis. Light is the welcome end to night that I feel each morning, as fearful humans always have. It is the long, pale, slanting rays of late afternoon sun in winter, conveying loss, nostalgia, an end to things. It is the rich crimson of a rose expressing sheer vitality as it vibrates against the crisp green of its bush. It is the sun's warmth on my skin, bringing a profound sense of inhabiting my proper place. It is the even expanse of a newly painted room or the uninterrupted blue of a tranquil sky, each calming my spirit. Something in my responses seems to go beyond the bald facts of wavelengths and biochemical reactions, yet physical reality underlies what I see: The blue of the sky is pure light of short wavelengths diverted from a straight path into my eyes, which respond as the energy of the light alters molecules in their retinas.

Laser beam and sunlight, physical effects and emotional, light ordinary and light exotic—but there is another way light enters my life. It is the medium caught and manipulated in the visual arts I love, to carry heightened reality to my eye and mind, to evoke response. Even without the element of color, the black-and-white photography of Ansel Adams or Edward Weston puts into play fine variations in light intensity. Gazing at a photograph, I see light reflected from its surface to re-create the velvety blacks, ambivalent grays, and shining whites left after photographer and camera have abstracted color from the scene, leaving behind pure form or feeling. Manipulated light appears again in the darkness of a movie theater. Strange to think that famous black-and-white moments like the opening tracking shot in Orson Welles's *The Magnificent Ambersons* or the final scene in *Casablanca* are noth-ing but moving patterns of intensities—bright here, brighter there, dim elsewhere—imposed on a beam of light by a strip of transparent film. In the monochrome world of early *film noir*, the *noir* is a view of life reflected in a dark palette, as in the minimal view of Burt Lancas-ter's waiting face in *The Killers.*

When color is added to still photography, cinematography, or paint-ing, the human response gains incalculable depth and breadth. Our eyes and brains can discern several million different shades. The range of response is impressive, but the pungency of color is clearer when we think about films. I recall how the world changed for Dorothy—and for me—when the stark black-and-white of Kansas gave way to the full-color land of Oz; I remember the pervasive golden glow in *Out of Africa,* and I see again the ravishing soft twilights in *Days of Heaven,* lit only by sky glow and hand lanterns. Light itself, called the true subject of photography, can also become a definite object in a film.

It is in the art of painting that light's color and intensity are best manipulated to evoke response. The choice of oils, acrylics, or water-colors; the selection of pigments and the mixing of tints; the use of thin washes applied with a brush, thick slabs of paint laid on with a spatula, or multiple planes of transparent and opaque color; the texture that comes from the painted surface and from brush strokes: All these shape the light that is reflected from a painting, which carries the artist's internal vision to the viewer. These methods create varied images and moods—the thick, whirling, explosive suns in Van Gogh's *Starry Night;* the flat tints that enhance the mystery of Henri Rousseau's *Sleeping Gypsy;* the deep, misty look of Claude Monet's *Rouen Cathedral;* the overwhelming blocks of color in Mark Rothko's art; the neon-vivid tropical shades of Paul Klee's *Fish Magic;* the use of light and shadow in Hopper's work to convey a sense of aloneness. And in religious art, light is employed to represent spiritual power.

As I step from the evocative world of art into the reality of the late twentieth century, light takes on another guise. Civilized societies have always sought the best possible artificial lighting. Now, the technology of light surpasses mere illumination. Carefully measured and controlled, its wavelength and duration set with exquisite care, light flashes beneath our city streets, carrying millions of voices and spates of digital data via optical fibers. Made by a tiny infrared laser within a compact disc player, it faithfully melts the music frozen in the disc into liquid sound. Rapidly displaying words, images, and numbers on a computer screen, light carries information at high speed. In medicine, it offers new ways to diagnose disease and to observe the living brain. Light in these uses is created or sensed by minute structures made of semiconducting materials, functioning at the atomic level, where quantum physics holds sway. This connection with the microscopic quantum world is another sign of light's universality.

In its applied guise, light defines our way of life, but there is a deeper connection. We could not exist without light, because all earthly life depends on solar radiation. Life is "woven out of air by light," wrote the nineteenth-century physiologist Jacob Moleschitt, poetically and accurately. What he wrote of was the process of photosynthesis, in which the sun's radiated energy is absorbed by the chlorophyll molecules that give plants their color. This initiates the chain of chemical reactions that changes water and atmospheric carbon dioxide into oxygen and carbohydrates, the stored energy of the plant. Humans cannot make this conversion. Instead, we survive on stored solar energy, by eating the plant that stored it or by eating the flesh of animals that eat plants. The photosynthetic harvest of light anchors the food chain that ascends to our own existence. Many processes of life, including our own apparatus for visual perception and the daily internal rhythms of our bodies, have evolved in response to the specifics of solar light.

Among the living processes influenced by light, surely the most significant are the marvelous workings of the brain that define what it means to be human. The network of billions of nerve cells and connecting strands that forms the brain has been called the most complex construction in the universe. A sizable fraction of that intricate web is devoted to light, beginning with its reception by minute sensors at the back of the eye and ending with its rapid analysis that gives meaning to a seen world full of shape and color. So pervasive is the visual function that it gives valuable clues to the behavior of the brain and even illuminates its deepest riddle, the evolution of human consciousness.

Ordinary sunshine or the glow of a light bulb; invisible emissions from a laser or from the far universe; manipulated energy that interacts with eye and brain to create visual art; tamed power for technology; life-giving solar radiation: All are light in its many facets. As a physicist I understand that different as they seem to human perception, all are aspects of the same phenomenon, variations on a constant theme: Light is a high-speed flow of undulating energy, a wave whose force lies in its crests and troughs.

The glowing green laser beam that pleases my eye, and the invisible ultraviolet ray that burns my skin, are both waves. They differ only in their wavelengths, the distance between two consecutive crests. For water waves, this can be feet or even miles. For light waves, it is measured in billionths of a meter. The shortest visible wavelengths produce our sensations of violet, indigo, or blue. Midrange values we sense as green and yellow, and the longest ones generate orange or red. Invisible light also comes in waves, but at wavelengths to which the human eye does not respond. Ultraviolet wavelengths lie below that of violet light, and infrared wavelengths exceed that of red light.

With their minute wavelengths, light waves are finer structures than are ripples in a pond or swells in an ocean. They are also less tangible. The ghostly hills and valleys of light's undulations represent varying electrical and magnetic effects, which are less easy to feel and visualize than the power of crashing water. But tangible or not, our own eyes tell us that these electromagnetic oscillations define the world we see. Or do they? Familiar as it is, light embodies a great irony and a profound enigma: Like nothing else that enters our ordinary world, it exists simultaneously as waves and particles. These particles are the crisp, bulletlike packets of energy called photons, first proposed by Albert Einstein in 1905. The energy that each photon carries depends on the wavelength of the light—the shorter the wavelength, the greater the energy.

This wave-particle duality is a puzzle at the core of physical understanding, but it also points to light's central role; its photons are one of the all-important types of elementary particles that have combined to form the universe. Physicists have found about two hundred kinds of elementary particles, but only a few are more or less familiar parts of the ordinary world. Protons and neutrons form atomic nuclei, which are surrounded by electrons to make the complete atoms that form the matter around us. We picture atoms in our mind's eye but rarely interact directly with their constituent particles. Other elementary particles are stranger, more distant. Some, like those called neutrinos, are prevalent in the universe but penetrate ordinary matter with virtually no effect. Without special methods, we would not even know they are there. Others, like the quarks, are made only at extreme energies in particle accelerators and exist only for the briefest eye blinks.

Among all these particles, photons are the only ones to which humans directly and regularly respond. Photons are also set apart from other elementary particles by their physical attributes. They do not coalesce into more intricate units as protons, neutrons, and electrons do. They are truly immaterial, in that they have no mass—the measure of the quantity of substance in a body. In addition, while the most exotic elementary particles live for millionths of a second at most, and even the garden-variety neutron separates into a proton and an electron under certain conditions, photons live forever. Light is truly the intangible, perfect particle. Is it any wonder it carries spiritual intimations?

Apart from its photons, light is fundamental in another way. Its enormous velocity in empty space of about 186,000 miles or 300,000 kilometers per second—a speed that covers one foot in a billionth of a second—amounts to a universal law, an absolute speed limit for every kind of moving body with mass, as established by Einstein's Special Theory of Relativity. Not only is this speed an inviolable limit, but it is a tenet of relativity theory: No matter how measured by any observer moving at any speed, the result for the speed of light is always the same. That constancy is an absolute feature of the universe, whereas the measured speed of any ordinary object depends on the speed of the observer. This strange property of light is integral to what is perhaps our most profound physical theory, one which changed our perceptions of space and time themselves.

Light displays yet one more fundamental aspect: In one particular form, it is direct evidence that the universe was jump-started billions of years ago. It was a revelation to me when I realized that the infrared light made in my laboratory is related to this pervasive light, this cosmic background radiation. How did light become the definitive signature of the Big Bang? To answer that question, we must go back to when the Big Bang made reality and to the photons with which we perceive that reality. It is irresistible to imagine the scene an instant before the Big Bang as dark—an absurd notion, of course, for light and dark, and even "before" and "after" themselves did not exist until the Bang created them. But the human mind is not equipped to envision "nothing" and can picture only featureless blackness instead. So imagine a dark and breathless moment some fifteen billion years ago, an instant of waiting, and then . . . the enormous eruption we call the Big Bang. All we know of matter—ships, shoes, sealing wax, cabbages and kings—comes from that one event, when energy and the very laws of nature were born. But the Bang did not make fully formed ships and kings. It produced the kindergarten blocks of the universe, so to speak, its elementary particles and the forces that were to form those blocks into galaxies and people.

While present knowledge does not take us back to the very moment of creation, we can follow events that came soon after, as established by a variety of scientific evidence and expressed in a dynamic scenario by Steven Weinberg in *The First Three Minutes*. At the age of a hundredth of a second, the universe blazed at a temperature beyond all imagining, a hundred billion degrees above absolute zero, compared to which the core of our sun is the merest ember. An enormous number of photons already existed, each carrying tremendous energy. Some of this light became matter, when the collision of two photons at these extreme temperatures would convert them into electrons and other elementary particles possessing mass. The infant cosmos was a dense stew of matter and radiation, a thick bubbling bouillabaisse of electrons, neutrinos, photons, protons, and neutrons. Like a ball of exploding gas, the universe grew relentlessly, its density and temperature decreasing as it expanded, and its mixture of elementary particles constantly changing.

A few seconds into the life of the cosmos, it had cooled thirtyfold to three billion degrees, too low a temperature to permit colliding photons to form new electrons. After three minutes, with the temperature at one billion degrees, our stew was spiced mostly with neutrinos and photons and still too hot to allow protons and neutrons to form atomic nuclei. But that event—the beginning of matter as we know it—happened soon after, when the temperature fell below a billion degrees. A proton and a neutron could then combine into the first type of atomic nucleus, the form of hydrogen called deuterium, and further reactions produced helium and lithium (the remaining hundred-odd elements of the periodic table that are now distributed throughout the cosmos were made much later, inside stars). Beginning about 300,000 years after hydrogen first appeared, when the temperature dropped to 3,000 degrees, these nuclei finally bonded with electrons to make complete atoms, the basis of common matter.

The incorporation of electrons into atoms explains the emergence of the infrared cosmic background light, the remnant of the photons in that early bouillabaisse of matter and energy. Before atoms formed, myriad freely swarming electrons impeded the motion of photons. Each collision of photon with electron stole energy from the light or changed its course. Light could not travel far; the cosmos was opaque.

Only after atoms entrapped those interfering electrons could light move freely. The universe became transparent as photons traveled its length and breadth. At 3,000 degrees, the energies carried by the photons corresponded to infrared wavelengths just beyond the present range of human sight. As the universe continued to cool, those energies diminished, and the wavelengths became longer in the inverse relationship characteristic of photons. The result is the cosmic background radiation that fills the universe, discovered in 1965. Its long, invisible wavelengths lie in the extreme infrared region, where my laboratory laser operates.

Scientists eagerly examine the infrared light in the heavens with sensitive receptors, to read more deeply in this encoded cosmic history. In the present epoch, the energies of the photons correspond to a cosmic temperature of about three degrees above absolute zero, a thousandfold cooling since light first freely traveled the universe. The calculable relation between the energy of the photons in the cosmic glow and a universal temperature changing over eons is strong evidence for the Big Bang theory. Researchers now measure the background light along different lines of sight into the cosmos, and they are finding variations that represent wrinkles in the early distribution of matter, which grew into the present heterogeneous pattern—immense clusters of galaxies separated by gulfs of space.

Particles of light are more than informative remnants of the early universe. They also help to bring the scattered elementary particles made by the Big Bang into complex assemblies of matter. Those fundamental particles resemble the plastic building blocks that children love to assemble, matching their bumps and hollows to form wondrous castles and airplanes. Similarly, the elementary particles are held together in intricate clusters by four types of fundamental forces that radiate among them: the nuclear force that binds closely packed protons and neutrons into atomic nuclei; the electromagnetic force, in which the long-distance power of electricity and magnetism links nuclei and electrons into complete atoms and atoms into the common matter around us; the so-called weak force that operates in certain radioactive processes; and gravity, which influences every kind of matter and holds together the largest structures of the universe.

Except for gravity, which despite three centuries of study from Isaac Newton to Albert Einstein is not fully understood, these forces can be explained within our general framework, for each is carried by a particular elementary particle. In this scheme, the significance of the photon is that it conveys the electromagnetic force. When light is taken as a wave, it is an electromagnetic wave that exerts electric and magnetic forces as it undulates. This feature remains when light is also visualized as particle, because photons produce all the effects we associate with electricity, magnetism, or both: They link the atoms of ordinary matter, generate electricity from giant rotating machinery, transmit radio and television signals, and so on.

These appearances show the human importance of light, even when it is invisible. And when it is visible, light carries aesthetic and emotional significance as well as the weight of its ancient place in the universe. I sometimes feel this fusion when I gaze at a work of art. There is my emotional response to the image, my physiological reaction that endows some colors with a vibrating intensity, my judgment of form and tint. Then my knowledge of the physical world brings thoughts of the complex interaction of light with matter, of how light strikes off pigments to set color and intensity, of the fundamental particles of light that convey meaning.

An examination of each of these faces of light is a progressive scientific inquiry that begins with light captured in the eye and brain, continues with light in the laboratory and the everyday world, and ends with light at the limits of the universe. Visual art provides a second thread of meaning shot through our progression, another approach to light that illuminates what science does not. Along each thread, we encounter many of the scientists and artists who understand light, and my own scientific involvement with it as well.

We know best the light we see directly. That anchors the beginning of our exploration—visible light and how we react to it. Thinkers from

Euclid and Descartes to twentieth-century scientists such as the Nobel Laureate George Wald have asked how humans sense light and give it meaning. Biological science now explains the process through the concerted response of rod and cone cells in the eye, which in turn is interpreted by neurons in the brain. But the interpretative stage is still not fully understood, nor is the influence of psychology and culture on vision. Art, too, shows us how we see. It provides tools to examine the complex manner in which the eye scans a scene. Study of the acute vision of artists gives further insight, especially in the work of Degas and others who transcended imperfect eyesight to see with artistic clarity.

The biological and artistic approach cannot serve for invisible light. Only physical analysis can combine what we see and what we cannot into one entity. The growth of that understanding is an ongoing scientific saga. The Greek philosophers of the fifth and fourth centuries B.C.E. began the serious contemplation of light. Near the end of the seventeenth century of the modern era, the astronomer Ole Roemer first measured the speed of light but did not know its constitution. In that same period, Isaac Newton analyzed light with a glass prism and came to interpret it as corpuscular. Later the wave theory became dominant, and in 1873 the physicist James Clerk Maxwell described light as an electromagnetic wave; for a time, that seemed the ultimate physical meaning of light. Then in 1887 the Michelson-Morley experiment changed everything, for it denied the existence of the ether, the medium thought essential to carry electromagnetic waves.

The dismissal of the ether was one factor in twentieth-century light. Another was the arrival of quantum theory, the idea that energy came in separate packets rather than a continuous flow, introduced by the physicist Max Planck when he analyzed light in 1900. The birth of the quantum, and the death of the ether, did more than create a new light that was both wave and photon. They created a twentieth-century physics, in which time, space, and gravity follow the theory of relativity, and light and matter display both wavelike and particlelike properties. Albert Einstein was central to these new understandings, and he was joined by Niels Bohr, Werner Heisenberg, and other innovative scientists of the early twentieth century. The search for understanding has continued through contemporary researchers, such as the Nobel Laureate Richard Feynman. The work of these scientists relates our present vision of light to modern physics, from Einstein's special relativity of 1905 to today's experiments that strive to clarify the dual nature of light.

The scientific search for classical light and new light influenced art, and that too is part of the story. Some artists paid homage to Newton by portraying the instant when his prism split light into colors, or by turning his view of the cosmos into art, as the imaginative French architect Etienne Louis Boullée proposed in 1784. In contrast, William Blake protested against the rationality that Newton had imposed on light, which Blake considered a limitation on the imagination. In this century, however, there has been an eager search for connections between modern physics and the new abstract art. Although there is no causal link, there are subtle associations. One of the founders of cubism, Georges Braque, was inspired to make aphorisms about science that serve to sum up the puzzling duality of particle and wave.

Even within that ambivalence, humanity has found pragmatic ways to make and control light, including the slow development of artificial illumination, from firelight to electric light, by the eighteenth-century chemist Ami Argand, who applied a new theory of combustion; the nineteenth-century technological entrepreneur Thomas Edison; and others. And when the power of quantum physics was brought to bear only three decades ago, the laser resulted. The originators of this marvelous light source—Charles Townes and Gordon Gould of Columbia University, Arthur Schawlow at Bell Laboratories—were rewarded with Nobel Prizes and valuable patents. Lasers are among the most powerful sources of man-made light, but even candles and military searchlights convey spiritual and political power.

We learned to control light after observing how matter modified light in the natural world, and these effects were explained and used, beginning with early thinkers like Claudius Ptolemy of Alexandria.

The law of refraction, definitively established in the seventeenth century, accounted for the bending of light rays as they passed from air to water and for the colors of the rainbow. Lord Rayleigh explained the blue of the sky in the nineteenth century. Such laws of light underlay the lenses and mirrors of microscopes and telescopes, that in the hands of Antonie van Leeuwenhoek, Galileo, and Newton first showed us the small and the distant—and still do. And in this century, we shape light in finer detail through the use of solid matter such as semiconducting materials that we understand at the quantum level. Now we can change photons into electrons and electrons into photons, linking light and electricity into a new technology for the next century.

The making and manipulation of light entered into art and changed how artists saw the world. Paleolithic artists could not have painted their caves without illumination from small oil lamps. In the sixteenth century, Caravaggio was said to have painted under artificial light to produce his dramatic effects. In 1912, Robert and Sonia Delaunay were struck by the new electric lamps lighting the streets of Paris, lamps which Sonia later portrayed as surrounded by colored halos. In recent times, James Turrell and Robert Irwin have made artificial light itself an artistic medium. And then there is the art and science of pigmentation. Van Gogh, Monet, and Henri Matisse could express colors blazing or delicate with a wide range of pigments, some of which also carry scientific meaning. One of the yellows beloved by Van Gogh contains cadmium sulfide, a semiconductor used in the modern science of light. Van Gogh's works themselves provide a kind of artistic laboratory of light as it is altered by natural effects.

Apart from artistic use, artificial light is most influential when it is built into complete systems. The most widespread example is the global web of optical fibers that carries telecommunications, and links computers via the Internet. A physically impressive example is the National Synchrotron Light Source on Long Island, New York, a device the size of a football field that has varied scientific and medical research clustered around it, including my own. Smaller systems of light animate novel types of computers, closely observe the thinking brain, and survey our planet and peer out at the cosmos from satellites in space. And technical light has contradictory impacts on visual art—validating original works, as in the recent attribution of the painting *Madonna with the Pinks* to Raphael, but also making it disturbingly simple to copy and alter works of art.

Our successive investigations have brought us to light at the edge of space, but not in the cosmos itself. Almost all our knowledge of the cosmos is carried to us by the light that fills space and shines from planets and galaxies. To engage that flow, humanity draws on each aspect of light—the visceral response to the beauty of the night sky, the deep-rooted need to find patterns among its stars, the technology of telescope and detector that gathers distant light, the physical understanding that relates light to internal processes of suns and black holes, to cosmic birth and death.

These facets of light have been brought together by the great stargazers. Van Gogh conveyed the spiritual power of the night sky and showed its astronomical features in his *Starry Night* and *Road with Cypress and Star.* As space travel began in this century, artists represented the universe in greater detail. In the 1950s, Chesley Bonestell painted scenes of cosmic beauty while taking care to portray them with scientific exactness. And then there are the scientists who gazed upward: Galileo, who in 1610 enthusiastically built a tiny handheld telescope to observe sun, moon, and stars; the ex-musician William Herschel, who discovered the planet Uranus in 1781 and later found infrared light in the sky; Edwin Hubble, who in the 1920s looked at distant galaxies for long hours and uncovered the red-shifted light which showed that the universe is expanding.

What these stargazers saw is explained by the modern view of light, which is embedded in modern physics. Einstein's relativity predicted the expanding universe, and in 1927 Georges Lemaître envisioned that explosion as starting from a minuscule seed—the origin of the Big Bang theory. Twenty years later came the theoretical prediction that the Bang would have filled the cosmos with a light that yet survives, the cosmic background radiation. And in 1965, Arno Penzias and Robert

Wilson found that light. Its role in the Big Bang is one of light's deep meanings, but not the only one. Light also holds a place in the Big Crunch, the cosmic contraction that could mean the death of the universe, or the beginning of its next cycle of expansion.

In returning to the birth of light, and perhaps its death, we close a circle. I picture the circle as the frame binding together the glowing segments of a stained-glass window, like the mighty rose window at Notre-Dame Cathedral. Each face of light we contemplate is part of a mosaic, its meaning enriched by its full shape. There is a matching scientific image also involving glass, but clear rather than stained, triangular in form rather than round—the prism that Newton used to refract white light into colored rays. After splitting light and analyzing it, Newton made a definitive test to determine that his understanding was consistent and true: He recombined the individual rays beyond the prism into a full beam of white light. Our faceted understanding is also reunited at the end of our search, into a full picture of light in the world.

2

Seeing Light

But man must light for man
The fires no other can,
And find in his own eye
Where the strange crossroads lie.

—David McCord, *Communion*

... being amazed and confounded by the supreme constructive ingenuity revealed ... in the retina [I felt pro-
foundly] the shuddering sensation of the unfathomable
mystery of life.

—Santiago Ramón y Cajal,
1906 Nobel Laureate for his studies of the retina

Seventeen thousand years ago, Paleolithic artists visualized the animals of their world as they painted their images on the walls of a cavern at what is now Lascaux, France. A century ago, Claude Monet
walked his garden at Giverny, and considered how to put its sparkling
color on canvas. In modern times, Ansel Adams contemplated Half
Dome in Yosemite, and Edward Hopper examined the Manhattan

restaurant that became the diner in his painting *Nighthawks.* Today, you or I peer through a camera lens to capture a lovely panorama or gaze idly at a scene. Ancient or modern, artist or ordinary viewer, each of us examines reality, or our memory of it, with our eyes.

Few of us can transmute what we see into a vision that moves others, as great artists do, but we all carry an intricate visual apparatus that had developed long before humans manipulated light in art or understood it through the intellect. We see the rainbow colors and a multitude of other shades, adjust to enormous variations in light intensity, and carry out prodigious feats of pattern recognition. What is clear, in the body's myriad visual sensors and nerve channels and in the masses of data they process, is that vision is our most complex sense—a stunning demonstration of the versatile power of living systems.

Great flexibility within changing visual circumstances is what characterizes the combined eye and mind, as part of its astonishing ability to impose pattern on the torrents of information carried by light. Look at an ordinary sheet of paper. Outdoors, whether under the sun's yellowish glare at midday or its weaker and redder light at sunset, you see the paper as white. Indoors, under illumination a hundred times weaker and probably bluer or redder than that of the sun, the sheet still looks white. A photograph of the paper, however, shows a rosy tint at dawn or dusk, as well as other differences that your visual processing assimilates but the camera cannot. The camera portrays exactly the scene that enters its lens, whereas the brain and eye together are like a color-correcting, auto-focusing camera that also knows to direct its own attention.

The brain and its visual functions are not yet fully understood. However, much of the eye's anatomy and its way of taking in a scene have been known since before light waves and photons were established scientifically. The structure of the eye and the marvelous lens that changes its focus for best vision had been accurately described by the first half of the seventeenth century. The great scientist, philosopher, and mathematician René Descartes studied the eye extensively. In 1637 he published in his *Dioptrique* a famous diagram that correctly showed light rays entering the lens at the front of the eye and brought to a focus on the back surface of the eyeball, the retina, to form an inverted image. This behavior had been confirmed in a striking, brutally direct experiment in which a scene viewed through the excised eye lens of an ox was seen to be inverted.

Since those early probes, it has been clear that simultaneous expeditions are needed to find the heart of vision. Psychology and physics, biochemistry, philosophy, and neural science—none gives complete answers by itself, because the study of perception must combine quantitative knowledge with qualitative and subjective aspects of our living response to light. Descartes joined science with philosophy to contemplate the process and meaning of vision. John Locke in the seventeenth century and Immanuel Kant in the eighteenth, along with other philosophers, continued to relate the internal process of perception to external reality. Perceptual psychology began in the late nineteenth century and continues to examine vision by studying how people respond to images and seeking meaningful patterns in their reactions. The art of pictorial representation, sometimes combined with psychological analysis, has also given insight into what it means to see and interpret.

These approaches probe vision as a whole. Another scientific view has developed in this century: what we see is determined by minute neural cells in the eye and brain that react to light and then organize those responses. This approach grows from advances in physiology made especially in the eighteenth and nineteenth centuries, when fine examination began to discern cells as the fundamental units of life. In the 1860s, microscopic study showed that the retina of the eye contains two different kinds of light-sensitive cells called rods and cones, after their shapes. In that same period, the physiologist and physicist Hermann von Helmholtz wrote clearly of light striking receptors in the eye and producing nerve impulses that traveled to the brain, where they were interpreted. This scheme explained in a general way how the inverted retinal image seen by Descartes—a result of elegant simplicity in the eye's optical design—appears correctly in final vision: The brain interprets the sensory impulses and turns the image right-side up. But the processes that converted light into neural signals and then analyzed them were not understood.

Twentieth-century biological science, operating on the microscopic level, explains how the rods and cones convert light into neural impulses and is beginning to grasp how these become a picture of the world. The complex story can be summarized through the work of a number of scientists who have contributed significantly in the last hundred years: Santiago Ramón y Cajal and Camillo Golgi, who made it possible to see neural cells through the microscope and showed that rods and cones are neurons connected into the intricate network leading to the brain; George Wald, who discovered how molecules of the substance rhodopsin sense light in the rods and initiate corresponding neural signals; and David Hubel and Torsten Wiesel, who together probed the brain with tiny electrical wires to determine how it deals with signals to interpret light.

These achievements in the modern understanding of vision are best understood by starting with a few basic facts about light—that it comes in waves and particles, which are described by wavelength, frequency, and energy. Those properties affected the eyes and minds of the cave artists at Lascaux, and they affect yours as you read these words, because they determine how and what we see. Wave theory nicely describes how the lens at the front of the eye focuses light on its retina, where sensing and initial processing take place. The colors we see are related to the frequency of light waves, as musical pitch is related to the frequency of sound waves. And photons also come into their own, for the act of vision begins as a molecule of rhodopsin changes its form in response to an impinging photon.

We have defined wavelength as the distance between two successive crests of a wave. Every kind of wave, whether swells in the ocean, sound in air, or light itself, also has a frequency that defines its temporal character. Frequency can be understood through Einstein's favorite method of exploring physical reality, a gedankenexperiment or thought experiment. Imagine yourself on the bank of a calm pond.

Drop a stone in the water and watch a lone circular crest spread out from the splash. Now lob stone after stone into the same spot at a regular rate, say one a second, to make an expanding set of concentric crests and troughs. A distant observer, unaware of your activity, could still infer that the force causing the waves repeated once per second, because that is how often a new crest would pass him. If you lob stones more often, the observer counts a greater number of crests per second. That is the frequency of the wave, measured in units of hertz after Heinrich Hertz, who confirmed the electromagnetic nature of light. Frequency has meaning for photons as well, for the greater the frequency, the greater the energy of the photon, in direct proportion.

There is also a connection between frequency and wavelength—the greater the one, the smaller the other (the mathematical relation is that the product of frequency and wavelength is equal to the speed of the wave). If you toss stones into the pond at a rapid rate—if the wave frequency is high—a given crest cannot travel far before the next one is launched. That means the crest-to-crest distance, or the wavelength, must be small. Visible light undulates at extremely high frequencies, hundreds of trillions of hertz. This gives minuscule wavelengths, whereas the sound waves that we can hear combine wavelengths of several feet with frequencies below 20,000 hertz. Such low vibrational rates can be created by mechanical structures, such as stereo speakers, and likewise sensed in the ear by mechanical means. But no conceivable gross mechanism can vibrate at the fantastic rate of trillions of hertz. In keeping with its ethereal nature, light requires more subtle means of generation and detection.

The ability to specify the wavelength and frequency of light introduces an objective order into the subjectivity of what we see. Consider the array of colors in a rainbow's arc, usually listed as red, orange, yellow, green, blue, indigo, and violet, working from the outermost circle of the arch toward its center. But that traditional set of seven colors is somewhat arbitrary. When Isaac Newton studied the colors in seminal experiments with glass prisms, he seems to have insisted on seven to match the musical notes A through G, in a belief that seeing and hearing were linked. There is evidence that he originally saw five colors, later adding orange and indigo. Some scientists who examine rainbows see six colors, excluding indigo; others have distinguished dozens of shades.

Although the colors seen in a rainbow depend on the observer, measurements reveal that they are arranged in strict order of wavelength, from the longest in the red to the shortest in the violet, and middle values in the green. A typical human eye responds with the sensation of sight for wavelengths from 400 to 750 nanometers, or billionths of a meter, from extreme violet to deep red. These wavelengths are comparable to a few thousand atoms set in a row; it would take several hundred of them to span the diameter of a human hair. The range of visible light may also be expressed in terms of frequency. Red light oscillates at 400 trillion hertz, and violet at 750 trillion hertz. In the alternate language of photons, whose energy is proportional to frequency, those in the violet carry nearly double the energy of those in the red.

A useful way to express the range is to borrow from musical notation, in which two notes are said to lie an octave apart if the frequency of one is double that of the other. Beginning with the musical A above middle C at 440 hertz, the A note one octave higher lies at 880 hertz, and the A two octaves higher is doubled again, to 1,760 hertz. The full piano keyboard represents nearly seven octaves; the range of human hearing—from 20 to 20,000 hertz—embraces nearly ten. Since the frequency of violet light is not quite double that of red light, visible light covers less than an octave. This is a narrow sweep compared to that of sound, yet we see a world full of astonishing color. Were our hearing limited to a single octave, say 4,000 to 8,000 hertz, music would become dull. We could not hear the deeper notes of a synthesizer or drum, the highest notes of a piccolo, and much more.

The richness of color seen in a mere octave comes at one of the eye's strange crossroads: its blending of physical stimulus and physiological response. Color is a sensation that light produces in the mind. The correlation between light of a single wavelength and perceived tint is direct; the brilliant emerald green laser beam in my laboratory has a single, precisely known wavelength. But generally we see by sunlight or by artificial light that contains a mixture of wavelengths, and then the eye responds in complex ways. Different blends of wavelengths can give the same perceived color, and some perceived colors, such as magenta, correspond to no single wavelength. And there is more: Any given hue may appear differently saturated (that is, with various amounts of white diluting the pure shade, as in pink relative to red), at different intensities, and set against other colors. Each variation changes perception, and experts in color reproduction estimate that we can discern up to ten million shades.

To consider only the abundance of color, however, is to miss another essential element of vision: its ability to cover space at high resolution. Even in a black-and-white world, we would process incredible amounts of data as the eye scanned a scene from side to side and near to far, absorbing its elements in great detail. Touch, taste, hearing, smell—each has its unique characteristics, but none traverses space with the power and ease of the gaze. We hear sound only from nearby sources; touch cannot range beyond the circle of arm's length; smell and taste come from intimate contact with foreign molecules. And these senses do not require the intensive mental processing that vision does. Smell is our oldest gateway to the world; its ingested molecules speak directly to the archaic core of the brain, which is why odors are so evocative. But when we see, only modulated energy enters the body, which must be decoded to describe the surrounding world. Vision trades the warmth of close response for the power to diminish space, yet this most remote of the senses can still affect emotion.

What we see may move us, and the varied movements of the eye determine what we see. A complicated set of muscles can rotate the eyeball in every direction, to smoothly track a moving object across the field of vision. But when the eye contemplates an entire scene, it does so by ceaselessly flicking over it in sudden movements called saccades, from the French for "jerk" or "twitch." These critical transits of the eye were discovered by L. E. Javal in 1878, who found them while studying the process of reading. We now know that a saccade is a dart of the eye from one position to another in a matter of milliseconds, followed by a fifth of a second or so at rest, then a repetition of the cycle. In short, vision is like a series of slides flashed at a rate of perhaps four per second.[1]

Saccades are symptomatic of intention in the act of seeing. Plato, Euclid, and other early thinkers speculated that the eye sent out rays to intercept an object of interest and return information about it. But the eye is no laser, responding only when an external source illuminates a scene. Nevertheless, we select what we see. The restless eye patrols a scene so as to bring to bear on its interesting elements the fovea, the part of the retina offering the clearest vision in its high resolution and sensitivity to color. We are generally unaware of these movements, but scientists have found clever ways to observe them. In the 1930s, the psychologist G. T. Buswell tempted the eye with art, when he examined how people look at pictures. In his experiments, a subject was seated so that a narrow beam of light was reflected from the front surface of his eye. As the subject looked at a picture, any movements of the eye changed the direction of the reflected beam, as if a small mirror were being turned this way and that. These variations were recorded on photographic film, to yield a map of how the gaze traveled over the image.[2]

One striking example of that moving gaze came when the subjects examined Hokusai's famous nineteenth-century woodblock print *The Great Wave off Kanagawa.* It portrays a mighty wave curling over at the top, its jagged fingers of spray clutching at two narrow shells crowded with rowers and nearly lost in the trough of the wave. Buswell's data clearly indicate the areas most frequently examined by the swings of the eye: the crest of the wave with its spray, its midsection, a smaller nearby billow, and the boats with Mt. Fuji in the distance. Least studied are the corners of sky and ocean far from the main wave. Buswell also found that consecutive moments of attention follow the curve of the wave—objective evidence that the eye "follows the line," as artists know. These results, along with modern ones, show that the combined eye and brain is not a copying machine mechanically scanning to make a faithful reproduction, but a selective apparatus driven by interest and intelligence.

The quick glimpses that underlie this selectivity impose a great complication, for they confront us with a world dismembered. It is only our cognitive power that holds the images in sequence, compensates for the fractured picture that results as the eyeball swings, and shapes it all into flowing reality. To accomplish this, the eyes must be closely connected to a brain that analyzes the information flowing from them. First, however, the eyes must sense light. The retina was taken early as the site of detection, imagined to be carried out by fine filaments that were somehow activated by light. But it took a breakthrough by the great Spanish neuroanatomist Santiago Ramón y Cajal near the end of the nineteenth century to clarify the essential nature of the sensors. He shared the 1906 Nobel Prize in Physiology or Medicine with Camillo Golgi, who developed a staining technique that enabled Ramón y Cajal to delineate individual nerve cells, or neurons, in the retinas of many animal species.

Before Golgi developed his method in the 1870s, a slice of brain tissue revealed little, for separate nerve cells were not clearly visible, even under the microscope. Golgi invented a chemical technique that preferentially dyed the elements of the nervous system with a form of silver that appears black. An image made by this "black reaction," still used today, shows a delicate tracery of individual neurons linked to one another through fine filaments that carry nerve impulses. Such pictures were a revelation in Golgi's time. They clearly established the brain as a set of discrete connected units, rather than a continuous medium. Today the image of an immense network of cells has become central to neural analysis.

As for vision, Ramón y Cajal's great contribution was to apply an improved Golgi method to the retina, showing that its cells are neurons as well. He made beautiful drawings of the cells as seen through the microscope that still form an atlas of their varied forms in different animals. We now know that the retina is a light-sensitive outgrowth of the brain, which early in human development is physically separated from the brain itself. Its sensitivity to light resides in the rods, which detect exceedingly dim light but are unaware of color, and in the cones, which sense color but require higher light levels. Both types are narrow cells, with lengths up to thirty times greater than their diameters of a few micrometers (millionths of a meter). More than one hundred million rods and cones cover the retina, like so many darts sticking out of a dart board. The end at the retina—the point of the dart, as it were—is where light is sensed when it reaches the back of the eye. The other end of each rod or cone lies near the optic nerves that carry impulses to the brain.

Both rods and cones appear in nearly every kind of fish, amphibian, reptile, bird, and mammal that scientists have examined, with variations that translate into differences in how each type of organism sees. In the human retina, six million cones determine our ability to perceive detail. The greatest spatial resolution occurs in the fovea ("pit" in Latin), a cuplike depression one hundredth of an inch across. It lies near the center of the retina, almost in line with the center of the pupil when the eye looks straight ahead. The fovea is densely packed with cones, like crayons nestled in a box. These are the slimmest receptors in the retina, only one micrometer in diameter, which allows thousands of them to fit into the equally minute fovea. Scaled to human dimensions, a similar cheek-by-jowl density would cram hundreds of people onto an area the size of a tennis court.

The intense concentration of cones falls off in any direction away from the fovea, until only rods remain at the extreme edges of the retina. What they do can be seen with your own eyes. Extend your right arm straight out horizontally. Turn your head to the left, then slowly bring it back to the right. You will find a point where the extended hand just appears at the rightmost edge of your vision, but only if you wiggle a finger, for edge perception is sensitive to motion; what does not move is not seen. Now hold a brightly colored object in that hand and, again finding the point at which you just perceive it, ask yourself to identify color. You know *something* is there when it moves, but try as you will, knowledge of its color is not to be had. This strangely split perception comes because rods sense motion but not color. To compensate, they operate at the absolute limit of sensitivity, for a single photon can activate a rod. However, it takes several such simultaneous activations to signal that a flash of light is present. This distinguishes true visual signals from false ones caused by random firings of the rods. Rods can detect the dimmest starlight, anchoring one end of an enormous range that extends a factor of a trillion to nearly full sunlight (although absolute full sunlight can damage them).

The only mechanism delicate enough to detect a single photon is a microscopic gate built into a single molecule of rhodopsin, which swings shut under the almost imperceptible impact of light. This elegant package is smaller and more effective than any sensors humans have devised. It is also parsimonious of ingenuity, in that the same method enables cones to realize color. Rhodopsin (from the Greek *opsis,* "sight") is a reddish-purple substance once called visual purple. A rod holds about a billion of its molecules, each containing a light-absorbing pigment called retinal whose actions start to convert light into neural signals.

The importance of retinal in human vision was found when the American biochemist George Wald followed a chain of evidence that began with vitamin A. This vitamin, like every other, is essential for normal metabolism, but is not made in the body; it must be obtained from foods. Those whose diet lacked the substance suffered night blindness, an impaired ability to see in dim light. (Yellow and orange vegetables are prime sources of vitamin A, the origin of the widespread belief that carrots are good for the eyes.) Wald discovered in 1933 that vitamin A exists in the retina and plays a central role in turning light energy into neural signals. In what he called "a quiet conversation with Nature," Wald studied the underlying chemical mechanisms for a quarter of a century, work that led to his Nobel Prize in 1967.[3]

Wald showed that retinal is formed in the body from vitamin A and explained how it detects light: A molecule of retinal is built around a long backbone of carbon atoms spaced like fence pickets, but with part of the fence bent, like a gate set slightly ajar. When the molecule absorbs a photon, the added energy changes the links among its atoms, bringing the gate into alignment with the fence. The straightening of the retinal molecule splits off part of the rhodopsin that contains it, and that freed portion goes on to accelerate a complex cascade of chemical reactions in the rod. These eventually release specialized molecules that travel across the gap from the rod to a nearby nerve fiber, producing an electrical impulse in the nerve—the first recognition by the neural network that photons have reached the retina.

Cones also detect photons, but with the key difference that cones come in three types. The light-sensing molecules in each are related to rhodopsin, with a pigment that responds differently to wavelength: one absorbs best in the blue (near 450 nanometers), one in the green-yellow (530 nanometers), and the third in the orange-red (565 nanometers). Just as in the rods, a shower of incoming photons generates nerve impulses, but these are now segregated according to wavelength. This filtration is what causes the sensation of color, for the brain interprets the relative numbers of the activated cone types as a particular shade. When light of wavelength 610 nanometers excites the retina, most of its absorption is in the red-orange pigment, with a small amount in the green-yellow and none in the blue; the perceived color is orange. Cone response also explains why the same color may be seen for different sets of wavelengths, if they happen to excite all three types in the same ratio. Conversely, some combinations of wavelengths make colors that can never come from a single wavelength, such as magenta, seen when blue light mixes with red.

Although the mechanism of color perception is understood, philosophers still disagree about the deep meanings of subjective response to color. Knowledge of cone response does, however, give a partial answer to the old question, How do I know that what you call red and what I call red are really the same color? Recent research shows that humans carry two versions of the orange-red responding pigment. About 60 percent of the male population has inherited one version, whose peak response lies five nanometers higher than that of the type in the remaining 40 percent. These groups indeed perceive color slightly differently.[4]

Interspecies differences in color perception are much larger than this. Some birds and fish have excellent color vision, but except for humans and other primates, mammals have only a rudimentary sense of color. These differences have presumably evolved in each species to enhance its survival, yet it is not known why color is so prominent in human vision. And why do we use the elaborate machinery of the saccades? Perhaps the combination of a small sensitive area with a clever aiming apparatus is the most economical plan, for it would be biologically costly to fill the entire retina with receptors spaced as closely as in the fovea. The neural capacity needed to aim and synchronize the swings of the eye is probably far less than that needed to deal with an entire retina working at high resolution. This intricate design shows the power of evolution to find subtle answers for living systems.

No matter how clever the design, vision makes extraordinary demands on the brain, which monitors a quarter of a billion sensors, coordinates data from two eyes, moves the eyes in a synchronized manner, and merges all this into a coherent picture of the world in real time. The task can be summarized in the language of computers, in which the unit of data called a byte is analogous to a word in human language. It takes forty million bytes to express the pictorial information in a single photographic color slide. Fifty or one hundred slides translated into bytes, as in the computer manipulation of images for printing, are enough to fill every cranny of data storage in a desktop computer. And for every second that the frames of a motion picture flash past your eyes, the contents of several computers are deposited in your brain, for you are seeing a billion bytes of information. Hearing is an undemanding exercise by comparison; in listening to digitally recorded stereoscopic sound, your ears take in data at a rate nearly two hundred times slower.

How is the mind to process it all? Part of the answer is that visual perception does not recognize each separate byte. Computers operate at that level of detail, but the eye and brain are far more sophisticated. They combine selective seeing via eye movement and direction of attention, early processing in the retina, and final massive analysis in the brain. Living intelligence imposes meaning that a computer cannot, to understand the visual field at a higher level of order than an uninspired flow of bytes. Otherwise, the thought of absorbing a stream of unmediated data at astronomical rates would be almost frightening.

Descartes realized that vision requires cognition at a level above what the eyes see. Aware that each eye is connected to the brain by an optic nerve, he concluded that the images from the two eyes are joined in a unified perception when the nerves meet at a single site in the brain. But the reality is more complex than that. The retina truly carries the analytical power of the brain beyond the armor of the skull, to provide the first step in visual processing. And although most of the work occurs in the brain itself, there too it is done at different sites rather than a single location. In his book *Consciousness Explained,* the cognitive theorist Daniel Dennett argues that consciousness is not something that happens at a central place in the brain, like the pineal gland that Descartes took as the seat of awareness; rather it is a set of events occurring at different times and places throughout the neural network. This idea of distributed cognition upsets our ingrained idea of self as a captain of being, located behind the forehead and commanding the body, yet visual perception does proceed in this decentralized way.

The presence of visual processing in the retina can be inferred from its complex neural arrangement. One set of connections goes directly to the brain. Like a vertical military organization, rods and cones speak up the line by connecting through units called retinal ganglion cells. Each of these is linked to one of the million nerve fibers, called axons, forming a massive conduit from each eye to the brain. Some fibers carry information from individual light receptors, especially those in the fovea, but the rods and cones are also cross-connected through other specialized cells, so that some axons carry the composite responses of many units. This added layer of organization performs the first step in mapping the outer world onto the neural one.

Descartes believed that the eyes transmit to the brain a reduced picture of the world. The neural map is, of course, not a picture; when you look at a tree, no neurons light up in your brain to mirror its shape, like Christmas lights following its outline. Rather, neural mapping redistributes visual information by attribute and location, resembling the strategy used to work a jigsaw puzzle. The puzzle box shows the final picture, say a red barn standing against fields and sky as cows graze nearby. But to analyze for assembly the thousand pieces jumbled in the box, the significant elements are sorted in nonpictorial ways. The simplest category, which even young children quickly grasp, is color and pattern: Into one pile go red pieces that come from the barn; into another, mottled brown-and-white pieces that are probably parts of cows. Geometric features form a second group: Any piece containing an artificial-looking straight line, for instance, probably shows an edge of the barn. A third classification considers location, so that a piece with a bit of blue sky or green grass belongs in the top or bottom half of the scene.

To comprehend the visual world, the brain rearranges it along lines of spatial location and visual property. The first sorting occurs at the retina. A typical ganglion cell senses illumination from several rods and cones arranged in a central bull's-eye and a concentric outer ring a small fraction of an inch in diameter. The ganglion cell fires, or is inhibited from firing, when one region is dark and the other illuminated. This antagonistic arrangement sharpens the neural reaction to changes in the contrast of illumination—significant because they often indicate the edges of objects. Changes happen in time as well as space, so the receptors also respond to motion, and some note wavelength. The outcome is that the retina immediately starts to analyze the scene impressed on it for edges, motion, and color.

The impulses from the ganglion cells enter the optic nerves that leave each retina. After further processing, the signals reach the visual cortex at the back of the brain, where the bulk of analysis occurs. The visual cortex is part of the greater cerebral cortex that covers much of the brain, the "gray matter" where most thought and perception occurs. Less than a quarter of an inch thick, the cerebral cortex is so heavily wrinkled that if unfolded, it would cover the surface of an office desk. Hundreds of millions of its neurons lie in the visual cortex (only one-tenth that number are committed to hearing), providing a hundred or more neurons to deal with each axon.

The visual cortex treats the incoming information in specific physical areas. The first the neural signals encounter is called V1, the primary visual cortex. Its neurons remap information from the bull's-eyes and rings of the retina, as if to further refine the piles of puzzle pieces. Some neurons respond to lines or edges tilted at particular angles. Others select for edges in motion or discriminate on the basis of color. This mapping is expressed in the elaborate spatial arrangements of the neurons. The nerve cells that sense a given angular orientation, say edges tilted at thirty degrees, form a vertical column that cuts through the six layers of the cortex. Adjacent stacks of neurons sense other orientations and intermingle the responses from the two eyes. A set of columns a fraction of an inch square represents all orientations of edges seen from both eyes for a small part of the visual field.

This detailed understanding of neural structure has been achieved by scientists who examine the visual cortex of an animal (usually a macaque monkey) one neuron at a time, using a device that David Hubel, who pioneered the studies with his colleague Torsten Wiesel, has called "the single most important tool in the modern era of neurophysiology."[5] This is the microelectrode, a minute metal wire with an extremely fine point, much smaller than a neuron. It is pushed into the area to be studied and brought near a neuron while the monkey looks at a screen displaying visual stimuli. As the neuron fires, the wire detects its signals and carries them into the laboratory. The signals are correlated with the visual stimuli to give a cell-by-cell map of visual function. Such delicate work by Hubel and Wiesel led to their discovery in 1959 of the columns that sense orientation—and to a shared

Nobel Prize in 1981. A newer method is to stain the neurons with dyes whose colors depend on the electrical state of the cells. This modern form of the Golgi stain produces vividly colored images of the highly ordered cortical arrangement.

Such ordered geometries are familiar to any scientist who, like me, studies the intricate atomic arrangements that make up solid crystals. Physics explains why some nonliving matter is crystalline, but why does the living brain map visual information into specific three-dimensional arrangements? One gain is in speed. Despite the phrase "quick as a thought," neural signals do not travel fast—from slower than a walk to a few hundred miles an hour. (The science writer Willy Ley once estimated that if you pulled the tail of a giant dachshund that stretched from Berlin to Bremen, two hundred miles away, the dachshund's head would not know to yelp until an hour later.) Brain cells that are laid out functionally speed cognition by minimizing the distances that neural impulses must traverse. Geometry also reduces the amount of genetic information needed to construct the brain, for a single "blueprint" describes many columns in the cortex. A more speculative possibility recognizes the plasticity of the brain, the changes in its neural connections that come from thinking. The placing of related cognitive activities near each other may induce useful relationships among neighboring neurons, like comparing an edge pile to a color pile in search of new clues for the jigsaw puzzle.

Visual processing does not end in V1, for the impulses proceed to other areas in the cortex labeled V2 through V5. Like V1, V2 contains precise spatial information; the other areas deal with color, form, and motion. One recent interpretation by the neurobiologist Semir Zeki of the University of London links V3 with moving forms, V4 with color and color combined with form, and V5 with motion. Knowledge of these areas is helping to clarify puzzling aspects of perception. In one such eerie phenomenon, called blindsight, damage to V1 can obliterate sight over much of the visual field, but some individuals so afflicted can still distinguish among visual inputs in their nominal blind spots. The startling feature is that they steadfastly insist they see nothing even while they make visual distinctions, as if removal of the V1 area strips away awareness while maintaining visual function. One possibility is that the precise spatial maps in V1 and V2 are necessary to comprehend results in V3, V4, and V5. Those areas return their findings to V1 and V2 to be properly located in space, as though jigsaw pieces that have been sorted into categories are finally interlocked into a picture.[6]

This speculation may not be correct, but it is certain that the decoding of light is spread over the neural network, and over time. And if visual processing is decentralized, so perhaps is visual comprehension. There is no evidence as yet for a highest "super-V" center that connects all the other V centers—the modern version of Descartes' locus of meaning. Such a coming together would resemble a central observer in the brain sitting back to understand the final picture, like a museumgoer pondering a photographic print. But if visual information sloshes back and forth among cortical areas, it may be that awareness grows as each pass adds a new layer of understanding. Thus visual comprehension is more like watching the image develop as a print sits in a photographic tray.

With its heavy use of neural resources, perception of light is so important a function of mind that it gives clues to reasoning and to consciousness. "Perhaps," says the perceptual psychologist Irvin Rock, "perception arose in evolution before conscious reasoning ability. . . . If so, we might conclude that thought is perceptionlike," not that perception is thoughtlike.[7] Francis Crick, who with James Watson discovered the structure of DNA, and his colleague Christof Koch study mammalian vision to illuminate human consciousness.[8] They hark back to the great founding psychologist William James, who believed that the process of consciousness requires attention and memory. The directed movement of the eye is a placing of attention, and there may be another form of selective attention, for there is evidence that specific neurons in a macaque monkey are activated by what interests the monkey in its visual field. Perception also requires memory; to interpret and recognize quickly, it must draw on stored information.

And once memory is involved, individuality and subjective reaction are not far behind.

So finally we stand at the strangest crossroads of the eye, where objective processes become subjective response; where impersonal quanta of light trigger neurons that elicit individual reactions. Light affects people, whether as pure color that influences mood or in the highly patterned form that is the seen world. Consider what it means to feel emotion when gazing at a photograph. The pathos of a starving child, say, is transformed into a picture and mapped again in the brain. Somewhere in that cerebral action, a set of buzzing neurons is felt as pity. But the links to vision are obscure; we do not yet know how the neural structures activated by light connect to sadness, or excitement, or other reactions of the whole person. Until those links are understood, broader strokes are needed. An examination of the human responses to light begins to untangle neurobiology from psychology, physiology from learned reactions.

The responses run from the universal to the particular, the spiritual to the mundane. Some enter into our oldest beliefs. Religion takes light as spiritual and moral center, its play against darkness as a morality play. The Bible, the Koran, the Upanishads liken the deity to light, or proclaim light the guide to nobility and knowledge. Sun worship and the gods of light have appeared from India to Africa, from British Druids to Mexican Aztecs, in Egypt, Greece, Rome, Scandinavia. On a less exalted level, there is the power of color in common experience. I notice how sunglasses affect my mood on a bright day. Glasses with dark green lenses cut glare but block out the warm colors, giving a gloomy cast that destroys pleasure under the sun. A second pair with brownish lenses makes the world warmly cheerful. But a different brown pair adds a yellow gloss, like certain photographic filters that enhance contrast and slice through the haze of distance. These glasses impose an unsettling yellow sky that seems to announce a tornado brewing just over the horizon. Is this reaction linked perhaps to memories of storms past, or is it a physiological response to certain wavelengths?

Color has definite and pronounced physiological effects. In *A Natural History of the Senses,* the poet and essayist Diane Ackerman relates how in New York's American Museum of Natural History she encountered a large piece of sulfur of so intense a yellow that she began to cry, expressing a direct pleasurable response from her nervous system.[9] And in fact to see the colors yellow or red raises blood pressure, respiration, and heart rate, whereas blue lowers them, giving a calming effect. Such visceral reactions may represent long-standing evolutionary patterns. Some primates prefer blue-green surroundings and are disturbed by red, perhaps because that color represents a potentially dangerous phenomenon in natural environments.

Evolution has brought about other strong connections between human response and sunlight. We see over the wavelengths at which the sun's output is most prodigious. Nearly half its radiation lies between 400 to 750 nanometers, and our eyes are most sensitive in the yellow-green that is the most intense solar color. We have direct bodily links to light, for light restarts the human clock every day, triggering a rhythm that dominates our lives. The body is designed to operate at a certain level during the daylight hours. It begins to do so when the impact of light on the retina activates a neural pathway to a part of the brain that controls alertness. People who live on unusual time schedules, or who quickly traverse time zones as they travel, suffer malaise as the rhythms of light are upset. The appropriate use of artificial light can be therapeutic for this problem and for the more serious affliction called seasonal affective disorder, the depressive state that some enter during dark winter months.

Physiological reactions to light also occur on less than a whole-body level. It has long been known that red elements of a scene seem to advance and blue ones to retreat, as used by the eighteenth-century French artist Jean-Baptiste-Siméon Chardin and noted in 1810 by Johann Wolfgang von Goethe in his *Theory of Color.* The reason is that as the lens of the eye varies its shape to properly focus the different wavelengths on the retina, the mind correlates the changes with apparent distance to the object. The effect can be exploited in works of art with red and blue areas purposely put adjacent. As the eye continually shifts its focus to accommodate each color, the flat image vibrates into the third dimension.

Apart from physiological responses, each color has its strong traditions: yellow, the life-giving sun and the tint of cowardice; blue, loyalty; white, purity. Alison Lurie discusses in *The Language of Clothes* what black sweaters, gray suits, and red dresses convey: Some shades stir physiological responses, perhaps sexual ones, as in the excitement of red. But other associations that color carries are set by culture, so that white, not black, is the color of mourning in some societies. Culture may even influence the perception of whole scenes. One well-known early study made among South African Bantus suggests that pictures drawn with European artistic conventions are interpreted differently by those who learn to see within another tradition, although this result has been questioned. On the other hand, the cave images at Lascaux and elsewhere, from times and cultures distant from ours, are outline drawings that we find understandable. Independent of culture, there seems to be something universal in how the mind interprets images and guides the hand to draw them in outline form.[10]

Whatever the role of culture in vision, one thing is certain: We each must learn how to interpret patterns of light on the retina. The neurologist Oliver Sacks has written the true account of a man, blind throughout much of his life, who regained his sight in middle age.[11] What seemed a miracle became a burden, for he found it wrenchingly difficult to understand the world through sight, rather than through the touch and hearing he had perfected for forty-five years. Even after considerable time, he was baffled by the changing appearance of moving objects; he sometimes could not differentiate his cat from his dog without touching them; he could make no sense of television or still pictures. Light showed him only an alien and impenetrable world, until finally he might have wished that his vision had not been restored, for the challenge of learning to see left him helpless and depressed. Such cases show, with other evidence, that only an elaborate process of learning that starts in our youngest days gives eye and brain the effortless interpretative power we later take for granted.

And we may learn to see better, by consciously incorporating new information. Scientists have always probed the world with their eyes, which looked deeper as they drew on mounting knowledge. Like Diane Ackerman, I enjoyed sheer color as expressed in the radiant gems and crystals I saw as a child in the American Museum of Natural History. Only much later did the accumulated scientific wisdom teach me that these pleasing visual properties are specifically related to the atomic details of solids. Now I look with informed eyes that give astonishing amounts of physical information: Transparent materials, like pure quartz, are brittle and conduct electricity poorly; those with metallic luster, like copper and gold, are malleable and conduct well; the rich colors of rubies and sapphires mark the presence of certain impurities added to a limpid compound of aluminum and oxygen. The beautiful appearance of these crystals shows the patterns of nature, to eyes trained to seek those patterns.

Knowledge, expectation, and emotion can make us reinterpret what we see. Some research suggests that motives influence perceptions; when a reward is offered for one of two figures that a subject may see, the one rewarded is more likely to be seen.[12] People reporting what they have observed as bystanders to a crime or in other stressful situations often give skewed descriptions. Certain optical illusions confront us with changing perceptions, such as staircases that shift suddenly from descending to ascending, a main element of M. C. Escher's 1953 lithograph *Concave and Convex*. Despite their training in careful observation, scientists are also influenced by illusions, as in the nonexistent canals of Mars reported by the astronomers Giovanni Schiaparelli in 1877 and Percival Lowell in 1894. Even apart from these striking effects, we carry our illusions, or at least our idiosyncratic vision, with us. Any nearsighted person can instantly change the look of the world from hard-edged to soft by removing his corrective lenses. In my own experience, magic happens if I let my myopia emerge at night.

Every street lamp, every neon sign and automobile headlamp is surrounded by a lovely halo. The less well I see, the more the world seems filled with light.

Paul Cézanne, Monet, and other artists have passed on to us a world interpreted through imperfect vision. Degas had a particularly serious set of difficulties. Author Richard Kendall, in his essay *Degas and the Contingency of Vision,* tells us that these were apparent not only late in Degas' life, as is often said, but much earlier. At the age of thirty-six, Degas found that he could not see a rifle target with his right eye; he long suffered from photophobia, or extreme sensitivity to bright light; parts of his visual field were blurred, and he developed a blind spot. The problems may be related to poor conditions during his military service at the Siege of Paris in 1870, which could have contributed to a retinal infection. Degas was also nearsighted, and later became farsighted as well.[13]

These deficiencies affected Degas' art during a visit to New Orleans in 1872, but not necessarily for the worse, as Kendall suggests. Though Degas was inspired by American scenes, his pictures from that trip are set in the interior lighting that was all he could tolerate, but in turn they are enriched by the inclusion of external light or outdoor views. As Degas confronted his difficulties, he may have become especially aware of the complexities of perception, thereby deepening his work. He believed also that the will is important in perception, saying, "Drawing isn't a matter of what you see, it's a question of what you can make other people see."[14] He may even have chosen to retain his nearsightedness. Photographs and pictures show that Degas almost never wore his corrective spectacles, which Kendall interprets as a purposeful decision to see a softer world. Cézanne also willed a personal vision by rejecting eyeglasses. The link between poor vision and valid art persists. There is evidence that some contemporary painters cannot see stereoscopically, that is, three-dimensionally. The perverse benefit is that this "handicap" makes it easier to compress the world onto a sheet of paper, ordinarily a difficult perceptual task.[15]

The tracing of perception through the eye, brain, and hand of an artist complements the study of the neural network. And yet, after tracking light into the eyes of the artists at Lascaux, onto Degas' suffering retina, into Ramón y Cajal's neurons and Wald's molecular gates, we could conclude that these impressive processes are in a way negligible, because we see the glory of light as through a narrow slit. The light-filled activities of the universe proceed for purposes other than our own. Some produce energetic X rays with short wavelengths of a few nanometers or less; others, like the cosmic background radiation, make wavelengths a billion times longer, deep into the infrared region and beyond. The light permeating the cosmos sweeps over forty or more octaves, and all but one of those octaves is invisible to us. Still, human explorers have carved footholds in the unseen: Johann Ritter, who first watched a crude forerunner of photographic film fade to black under ultraviolet light; William Herschel, who found the infrared in sunlight; Wilhelm Röntgen, who made X rays and showed an amazed world the skeleton of a living hand; Heinrich Hertz, the first to create and detect radio waves.

Without producing the sensation of sight, these invisible rays affect us, sometimes harmfully. Emotional associations with blood and anger might suggest that danger lies in the red, but actually it is just past the violet where care is needed. The poetic power of the "blues" has a real basis, for the blue-violet end of the spectrum is the portal to powerful ultraviolet and X-ray radiation, which can kill. X rays have frequencies above ten thousand trillion hertz. The energy of a photon rises with its frequency, and X-ray photons are energetic missiles that can smash links between atoms or break loose their electrons, to destroy the delicate chemistry of life. Ultraviolet photons lie between 4 and 400 nanometers, but only those from 200 to 400 nanometers play a significant human role: They are less energetic than X rays but may be even more harmful.

Ultraviolet light comes from the source responsible for earthly life, the sun, which emits one-tenth of its light at wavelengths below 400 nanometers. That radiation is diminished as it passes through our atmosphere, where ozone, a molecule of oxygen made of three atoms rather than the usual two, absorbs light below 295 nanometers. This shield becomes less effective as man-made chlorofluorocarbons erode the ozone layer. But even if chlorofluorocarbons had never been used, ultraviolet light would still reach the earth's surface in harmful amounts. These photons can damage human DNA, our built-in blueprint of cellular replication. If so, a cell will not clone itself identically, and cancerous growth may result. Statistical evidence shows that solar light in the so-called UVB region—280 to 320 nanometers—is implicated in skin cancer, now the most prevalent type of cancer. There is also evidence that ultraviolet light weakens the body's immune responses, the natural limiters of damage that prevent further growth by a sick cell until it has healed. After ultraviolet missiles disrupt cells, the greater devastation may be that they prevent the body from successfully containing these early injuries.

As if to compensate, ultraviolet light also treats cancer. In one method, the patient's blood is removed, exposed to ultraviolet rays, and returned to the body. This relieves symptoms of a certain kind of skin cancer, without showing the side effects that accompany chemotherapy. The method also shows promise for some forms of leukemia and has been used to treat a small number of AIDS patients. Ultraviolet radiation may also activate certain drugs that are sensitive to light, as in the treatment of psoriasis, the appearance of scaly patches on the skin.

The more powerful X rays show the same mixture of good and ill. Their devastating effects on living tissue were not understood in the first years after Röntgen's discovery in 1895, and many experimenters and patients suffered unnecessary harm. Now X rays are essential for medical diagnosis, but great care is taken to minimize exposure times for both operators and patients. Like ultraviolet rays, X rays also treat some cancers, whose aberrant cells they can sometimes selectively destroy. At the other end of the spectrum, the low frequencies beyond visible red light are associated with less energetic, less harmful photons. Infrared light is felt only as warmth on the skin, although in excess or over long exposure it can burn. High-power microwaves, with wavelengths of several centimeters, can also burn. Radio waves of very long wavelengths are expected to create no harm, although some claim to observe subtle changes in the nervous system.

What if we could see the invisible light that presently lies outside the visible range of 400 to 750 nanometers? Suppose human vision responded over ten octaves, the same factor that hearing covers, but began where our sight presently ends, at 750 nanometers, and extended to 750 micrometers. We would comprehend reality very differently, for we would see a luminous infrared universe. The sun would still look supremely bright, emitting as much light over the ten octaves of the infrared as over the single octave of the visible, but the nature of darkness would change. According to the laws of physical light, all hot bodies emit a range of wavelengths. Extremely hot sources shine mostly in the visible range, whereas cooler ones glow in the infrared. In our imagined world, anything standing a little above ambient temperature would take on its own visual life after sundown. A rock that had baked under the sun, the working engine of an automobile or airplane, our own bodies—each would show its presence. Perhaps night would seem less threatening, and less magical. The cosmic background radiation might appear as a dim glow in the night sky. And the map of the skies would change, for each planet, star, and galaxy radiates differently in the infrared than it does in the visible.

In the human world, we might be able to see bodily tumors, whose high metabolic rates change their temperatures and hence their infrared emissions. The hidden links woven through our technology would appear, as the infrared beam generated by every video remote control unit sprang into view. The rich array of new colors that we might see over ten octaves seems impossible to imagine, but the writer Alfred Bester has at least given them evocative names. In his classic science-fiction novel *The Stars My Destination,* a character with infrared vision sees a color spectrum that ranges over "deep tang to brilliant burn."[16]

The range of our imagined vision could also cover ten octaves in the other direction, from 0.4 nanometers to 400 nanometers. This would be a dark world even after sunrise, for the ultraviolet and X-ray light from the sun is only a fraction of its visible light. But the night sky would be enlivened by certain incredibly intense cosmic sources. There is Sco X-1, for instance, in the constellation Scorpius; if the X-ray energy it emits in one second could be stored, it would provide electrical power to the entire United States for a billion years. On Earth, wherever X rays could be found, the eye would see them shining through heretofore opaque materials—metal, plastic, our own flesh—to reveal what lies within. Ultraviolet light would show hidden features of the living world, such as patterns on flowers that may attract the attentions of bees, who see wavelengths as short as 300 nanometers. And the tiny ultraviolet and X-ray wavelengths are affected by small physical structures, so perhaps we could see individual atoms and molecules in a new direct vision of the microscopic world.

My visions of imaginary worlds illuminated by invisible light are only fantasies to show how light entwines with perception. For the light between 400 and 750 nanometers that we actually see, the reality is that its perception is now understood at the fundamental level of discrete receptors and neurons. The next stage of organization—neurons linked into columns and "V" areas—is also yielding to study. A different level of understanding, with less detail but greater sweep, comes from psychology, anthropology, religion, visual art. These dual explorations, one starting from neural analysis, one beginning with the whole human, promise to meet some day, but their convergence is still far distant. That final comprehension of the complete visual human seems essential to understand how we think and feel.

Such comprehension would still not grasp the essential physical nature of visible light, let alone the light we cannot see. Dual explorations have been mounted there as well, for as physical understanding grew, the particle picture and the wave picture of light each had its adherents. By the end of the nineteenth century, scientific consensus had been reached: All light, visible and invisible, was a wave that carried electric and magnetic fields. That classical view has given way to a confused new image of light in this century. Now it seems that light is a wave when it travels through empty space, but it becomes a multitude of particles when it encounters a solid or is made in a laser. Each model is valid in its sphere, but we have yet to unite the two. How scientists first struggled to choose between wave and particle, and now struggle to join wave with particle—the great puzzle of modern physics—is what we examine next.

3

Classical Light

Nature and Nature's laws lay hid in night:
God said, Let Newton be! and all was light.

—Alexander Pope, *Epitaph for Sir Isaac Newton*

Those early ancestors of ours, seeking to understand the physical world, had to grasp both its matter and its energy. Matter, whose gross properties could be felt directly, was more comprehensible than light. A scene in Stanley Kubrick's classic film *2001* helps to make this point when it shows one of our prehistoric ancestors rooting among a heap of bones; puzzled, he peers at a large bone, smells it, then picks it up and holds it. Inspiration seems to follow as he lifts up the bone and brings it down forcefully to shatter a skull lying nearby. It is not long before his tribe is beating off invaders with heavy clubs modeled from bones. Matter, in Kubrick's example, was densely *there*. When humans hefted a club or shaped a piece of flint, each was experiencing its material attributes such as inertia and hardness. Light, by comparison, was elusive and exotic. Pre-scientific folk could neither heft nor manipulate the glow of a fire or moving glints of sunlight reflected from the sea; they were totally unaware of the characteristics of invisible light.

Only much later came quantitative methods to examine the

subtleties of light, visible or invisible. Thousands of years after artists first made images from nature, our understanding had matured sufficiently to create models of light based on the material world. We could imagine discrete particles of light that behaved like rocks hurled through space. Or we could visualize light as a wave, its fundamental features resembling the rhythms of the sea. Both models—localized particles each delivering its single impulse, or extended waves passing through a supporting medium—soon were woven into the historical understanding of light. The methods used to distinguish between the two pictures could fill a scrapbook full of snapshots showing how physical knowledge grew (and continues to grow), in the interaction of theory and experiment, mechanical model and mathematical description.

Today we know that light is more tangible than it first seemed. It exerts physical force, an effect that has inspired visions of spaceships with enormous sails pushed by sunlight. In Einstein's theory of relativity, light has weight of a sort, for it is affected by gravity. Nonetheless we do not yet have a unified understanding of light. The wave picture describes light moving through space like an ocean comber rolling on far from land. Photons are invoked when light interacts with matter, as if a wave of light were breaking on the shores of something material to become a spray of quanta. But how does one become the other? How do separate photons correlate their actions to make a single wave? What is the meaning of a single, distinct photon? Similar questions enter into the quantum mechanics of matter, whose primal bits also have an extended, wave-like character. Until these puzzles are resolved, our comprehension of light and of matter remains incomplete.

And beyond the tension between the two models, there is the question of the speed of light. Even within a universe forever in motion, all matter—from elementary particles to galaxies—comes to rest, or nearly so, at the temperature of absolute zero. But like a shark that must swim or die, light in space can never rest. If it exists, it moves, at 186,000 miles per second. That enormous speed in vacuum is an absolute limit for any imaginable motion. It is attained only by light itself, or possibly by the elementary particle called the neutrino.

Physical light behaves like nothing else we know. Moreover, our acute perception brings it to us loaded with subjective meaning. To some extent, light must be seen from inside our own vision, just as we struggle to understand our consciousness from within its own limits. Some thinkers, questioning whether we can apprehend light in a purely objective way, have favored sensory knowledge or spiritual awareness. Newton was hailed for his scientific study of light, yet von Goethe opposed Newton's experimental approach, espousing instead a theory of color based on perceptual experience. The visionary William Blake went even further, decrying the rational dismemberment of light as an unspiritual act.

Though the ancient Greeks seriously considered light, they encountered it only through vision, which made it difficult to separate its objective attributes from the subjective act of perception—one reason why early uses of light did not quickly reveal its nature. Mirrors are mentioned in the Biblical book of Exodus, written before 1000 B.C.E., and some have survived from Egyptian tombs nearly two millennia old. These were made of polished metals such as bronze, probably the first controlled interaction of light with matter. The properties of lenses were appreciated as well. Aristophanes mentions a burning glass in his play *The Clouds,* and the Romans knew that lenses magnified images. And some of the natural manifestations of light were codified. In 300 B.C.E., the book *Optics,* which was attributed to the great mathematician Euclid who wrote the classic *Elements* of plane geometry, correctly assumed that light travels in straight lines. Plato noted the distortion of an object as it extended from air into water, now ascribed to the refraction or bending of rays of light, and around 100 A.D. Claudius Ptolemy of Alexandria examined the effect in more detail.[1]

These studies formed the basis of geometric optics, which uses linear rays of light to analyze lenses and mirrors, but did not explain the composition of the rays. The Greek philosophers of the fifth and fourth centuries B.C.E. speculated about light as they considered vision. Pythagoras of Samos and his followers; Empedocles, who postulated a cosmos made of earth, air, fire, and water; Democritus, who espoused a world made of atoms; Plato; Aristotle—each theorized about light and vision, according to his worldview. Some of these thoughts are scientifically valid, some are misleading. The Pythagoreans believed a "visual ray" proceeded from eye to object, the inverse of what actually happens. The charismatic figure Empedocles, born in Sicily in the fifth century B.C.E. who, it is said, died by leaping into the crater at Mount Etna, also believed that light emanated from the eye. He related the process to his four elements: The eye is like a lantern, he wrote, containing an internal fire that passes outward through its water-filled portion to illuminate the world. But he seems to have believed that simultaneous illumination by sunlight was also necessary for vision. Plato shared this view, an early inkling of the importance of external light.

There also appeared suggestions of light as something that traveled or resided in a medium. Democritus, who hinted at the idea of a medium of transmission, believed that a seen object emits particles; he thought the emission imparted to the air an image impressed on the soft, moist eye. Aristotle criticized the extant models of vision, including the idea of a visual ray, arguing that if the eye is like a lantern, we should be able to see at night. He proposed instead that vision occurred when a light source changed the condition of an intervening medium from opaque to transparent, an idea that did not clearly explain how the alteration came about. The historian of light Vasco Ronchi poignantly suggests that perhaps Aristotle lacked the words to express his seminal insights.[2]

The ideas of the Greeks continued to carry weight in Roman times. In his great poem *De Rerum Natura* (*On the Nature of Things*), Titus Lucretius Carus presented to Romans a view of the world based on Greek Epicurean philosophy and the atomism of Democritus. Lucretius addressed light and vision at length, relating them to the properties of atoms. But after the Roman era, early conceptions of light were not much pursued in Europe. However, one of the great scientific scholars of the Arab world, the Egyptian Ibn al-Haytham or Alhazen, preserved and extended such studies around 1000 A.D. He examined how light is reflected, drew the first detailed picture of the eye, and gave compelling reasons why vision must proceed by external light (for instance, how could a ray leaving the eye account for the pain felt when looking at a bright light?).[3] His work, translated into Latin, created a spurt of optical activity in medieval Europe. The best known among those he influenced was the thirteenth-century Franciscan friar Roger Bacon, who extended our understanding of lenses.

After Bacon's death, there was little additional investigation of light until the seventeenth century, when scientists began to separate the essence of light from its applications, its geometric behavior, and its visual effects. Descartes exemplified this approach when in 1637 he formulated a physical model of light, described as a pressure transmitted through the *plenum*, a fluidlike substance that filled every void in the universe. The model was consistent with one important outcome of the early considerations of vision: The Greek philosophers noted that as we open our eyes we see near objects and distant ones, even those to the farthest limits, within the same instant. It seemed natural to think that light must travel extremely fast. Some even believed its transmission to be instantaneous; that is, to occur at infinite velocity. Descartes was a strong proponent of this view, and he made infinite speed a central feature of light in the plenum.

Speculations about the speed of light could be resolved only by physical measurement, one of a long line of experiments whose cumulative results would come to portray light. Whatever light was, its velocity could be found by determining the time it needed to cover a known distance. Although we now know that light moves too fast to track with just any timepiece, the experiment was proposed in the seventeenth century by Galileo, a founding figure of experimental physics. I encountered a variant of Galileo's method on the way to my doctorate, when an examiner asked me to consider its meaning. In the experiment, a student of light is dispatched to the top of a hill. He carries a lantern with a shutter, which he opens at a designated time. A second researcher, let us say Galileo himself, stands on a hill five miles distant with a synchronized timepiece. He records how long after the opening of the shutter he sees the light from the lantern. What, I was asked, could Galileo conclude from this?

Any modern physicist can calculate that light flies over a five-mile course in just twenty-seven microseconds—too short a time to allow Galileo to respond or to be measured by a seventeenth-century clock. So the narrow answer to the examiner's question is that Galileo's experiment would have failed. My interrogator, however, expected a reply that cut closer to the heart of the matter: As Galileo realized that the light had arrived before he could even register the fact, he would appreciate that light moved "damned fast" (as it was put to me), certainly much faster than sound. If the experiment were to use the blare of a trumpet rather than the flash of a lantern, the elapsed interval would be nearly a million times longer, twenty-four seconds. Although "damned fast" is fascinating information, it does not distinguish between "infinite" and "extremely rapid but finite."

If early measurements could not deal with light flashing over earthly distances, the alternative was to work on a bigger scale. The first to do so was the seventeenth-century Danish astronomer Ole Roemer, whose laboratory encompassed the impressive vault of space out to the orbit of the planet Jupiter. For nearly a decade he had been observing the Jovian moons that Galileo had discovered through his telescope in 1610. The work was motivated by the search for an accurate astronomical clock that sailors could use to set their position at sea. Roemer measured the time it took the moon Io to swing once around its planet, by noting the intervals between its successive eclipses in Jupiter's shadow. In 1676 he announced that the period, expected to be constant (just as our own moon orbits the Earth with a period of some twenty-nine days), instead became shorter or longer as Jupiter and the Earth neared each other or drew apart, respectively. The discrepancy was as large as twenty-two minutes for observations made six months apart, when the Earth had moved halfway around its orbit. In a leap of logic, Roemer concluded that he was measuring the sum of the actual orbital period and the time needed for light to speed from Io to his eye. Light, he inferred, had a finite velocity.

Roemer's data allowed Christian Huygens to calculate that velocity two years later. This productive Dutch scientist discovered the rings of Saturn, is the probable inventor of the pendulum clock, and developed a wave theory of light in 1690. Combining Roemer's time discrepancies with the distance traveled by Io's light, Huygens announced that the speed of light was 48,000 leagues per second, or in modern terms, 144,000 miles per second. The result differs from the accurate value of 186,000 miles per second because the distance was incorrectly estimated, but it is of the right magnitude, and most important, it is finite.

Descartes would have been chagrined to see this result smash the validity of his idea of the plenum. In fact the finite speed was one reason Huygens favored a wave theory for light,[4] and it has another profound consequence: It makes light a probe that can take deep soundings of time, back to the most ancient cosmos. If light moved from place to place with no elapsed interval, we would see only the present moment throughout the universe. Photons from a distant star, and those from a nearby candle, would represent the same instant of "now" at each source. But because of the finite speed, to see light from far away is to peer down a tunnel in time anchored in earlier events. The unit of astronomical distance called the light year—the six trillion miles that light travels in a year—also represents travel in time; when we observe a star a thousand light years away, we see what it was a millennium ago. When the Earth lay at its nearest distance to Jupiter, 368 million miles, Roemer watched an eclipse that was already thirty-three minutes old. And now that electronic technology can split time into tiny shards, the finite speed of light is meaningful on earthly scales. In my research I create brief pulses of blue or green laser light that alter solid materials as they are absorbed. I watch the laser beam travel its intricate path on my laboratory table, and I note that the distance of twelve inches is almost exactly a light nanosecond. Even that eyeblink of a billionth of a second to travel a foot affects my experiment and limits the speed of some electronic devices.

Science thrives on numbers, and efforts continued to measure the speed of light to greater accuracy. Such extreme certainty can seem mere dull fact, but sheer number sometimes takes on surprising life. Precise knowledge of how fast light moved would later point to its electromagnetic nature and to its role in Einstein's relativity. In the first valid measurement on the Earth's surface, Armand Hippolyte Louis Fizeau cleverly completed Galileo's experiment. In 1849, in the Parisian suburbs of Suresnes and Montmartre, Fizeau arranged a light source and a mirror slightly more than five miles apart. He timed the 58-microsecond round-trip with a rotating wheel whose rim carried many gearlike teeth. Light from the source passed through the gap between two teeth when the wheel was at rest, continued to the distant mirror, and rebounded back through the gap to Fizeau's eye. With the wheel turning, careful adjustment of its speed set the timing so that the returning beam encountered either a tooth, when Fizeau saw no light, or a gap, giving full intensity. Relating these changes to speed of rotation and distance traveled, Fizeau found a velocity of 195,000 miles per second, a figure off by only 5 percent. Later techniques gave higher accuracy until the 1870s, when the young American physicist Albert Michelson measured nearly the exact modern value—186,282 miles per second—with methods that would reappear in the famous Michelson-Morley experiment.

When Roemer uncovered light's finite speed, he eliminated the theory of the plenum but did not need to consider whether the images he saw were carried in from Jupiter's orbit by vibrations or by particles. It took a different line of development to notice that light had both its undulatory aspects and its particle-like aspects. Because it is the ambiguous shift between wave and particle that complicates our grasp of light, it is essential to make the clearest possible distinction between them. If multiple missiles assail a target, from cannonballs battering a castle wall to raindrops striking a window, the effect is always greater than that from a single particle. But two or more waves can be either stronger or weaker than one, depending on how they are brought together.

Two waves are said to interfere if they overlap in space. Their total impact is set by how the peaks and valleys of one wave stand relative to those of the other. Suppose the crest of an ocean wave towers ten feet above the sea at rest, and the following trough dips ten feet below. If two such waves interfere so that their maxima and minima coincide, they add up to a single stronger twenty-foot wave. That outcome is called constructive interference. But if the peaks of one wave lie just at the troughs of the other, the interference is said to be destructive. Each up or down displacement of water due to one wave is exactly canceled by the opposite excursion in the other, leaving an undisturbed ocean. This complete cancellation is used in sound-deadening headphones made for those who deal with noisy environments. The device analyzes the sound waves from the noise and rapidly generates countervailing waves to silence them in the listener's ears.

It is a critical difference that interference occurs for waves but not particles. Destructive interference between waves could explain why two overlapping light beams produce dark regions on a screen as well as bright ones. It could account for diffraction, the bending of light around an obstacle that frames the resulting shadow with alternating dim and bright bands. Interference could also explain the colors seen in the delicate film of a soap bubble. Natural light of many wavelengths is reflected from the front surface of the film. Part of the incoming light continues through the film, is reflected from its rear surface, and re-emerges through the front surface. As it interferes with the light originally reflected there, some wavelengths combine destructively and disappear, leaving a fine shimmer of color that changes with the thickness of the film. Seventeenth-century researchers associated these aspects of light with wavelike features. Francesco Maria Grimaldi, professor of philosophy at the Jesuit College in Bologna, thought diffraction effects were ripples resulting from a fluidlike light flowing around obstacles. Robert Hooke, of the British Royal Society for the Improvement of Natural Knowledge, also studied diffraction and the colors of soap bubbles. His *Micrographia* of 1665 suggested that light was a fast-moving, rapidly vibrating phenomenon.

But the wave model began to be eclipsed in 1666, when Newton performed the best-known experiment in the history of light: He inserted a glass prism into a sunbeam and separated its apparent seamless whiteness into a rainbow of colors. Many consider Newton to be the greatest scientist of all time, a reputation that relies heavily on his superb analytical talents. He was Lucasian Professor of Mathematics at Cambridge University, the position now held by the theoretical physicist Stephen Hawking. Self-taught, working alone with only mind, paper, and pen, Newton invented the core scientific tool of differential calculus and developed theories of mechanics and gravity that explain earthly and celestial motion, the ocean tides, and more; these endure yet, guaranteeing his immortality. Along with his acute theoretical insight, Newton probed reality with significant experiments and knew how to extend initial results into more decisive measurements. These were aided by his skill as a hands-on craftsman, which stemmed from ardent mechanical tinkering as a youth.[5]

His combined analytical and experimental talents gave Newton unrivaled scientific power. Things have changed in present-day research, in which few theoretical scientists perform experiments or make devices. Given the sprawl of contemporary science and the cost and complexity of its research equipment, this seems unavoidable. The modern scientific method that plays theory and experiment against each other until concept and testable result converge, however, mirrors Newton's insistence that theory be firmly connected to experiment. The best procedure for "philosophizing," he wrote, is "first to search carefully for the properties of things, establishing them by experiments, and then more warily to assert any explanatory hypotheses."[6]

Newton combined the conceptual with the actual in the novel optical device that brought him to the attention of the Royal Society. In 1669, he designed and built the first reflecting telescope, of which he was inordinately proud.[7] It used curved mirrors rather than lenses to magnify distant scenes, without distorting their colors as lenses did. His first model, only six inches long and an inch in diameter, magnified by a factor of thirty. The design is so advantageous that today's largest research telescopes still use mirrors rather than lenses. Soon after, Newton described his other optical studies to the Royal Society. He had unraveled light with a glass prism of triangular cross-section (the first of several that he used was purchased at the Sturbridge Fair) set in a beam of sunlight entering a darkened room.[8] The experiment was arranged so that light leaving the prism traveled some distance before falling on a wall, ensuring that the separate rays of light diverged sufficiently to show clear bands of color there.

What split the colors in the prism was the process of refraction (Newton called it refrangibility), which causes light rays to bend as they move from one medium, such as air, to another, such as glass or water. This is what makes a stick partly immersed in water look bent to an observer above the water's surface. Newton believed that white light contained separate rays of color, each bent through a different angle at the interfaces between air and glass prism. He was admirably clear on this point, writing: "To the same degree of refrangibility ever belongs the same color, and to the same color ever belongs the same degree of refrangibility. The least refrangible rays are all disposed to exhibit a red color . . . the most refrangible rays are all disposed to exhibit a deep violet color. . . ."[9] Newton's first observations were followed by sharper confirming experiments that recombined the colors into white light, proving that the effects he saw were indeed due to refraction.

Newton's measurements reinvented white light as the sum of simpler entities. As to the essence of those basic units, he weighed different ideas. Newton spoke of rays of color but also of particles or corpuscles of light, in keeping with his sense of a mechanical universe, and of vibrations induced by the corpuscles in a pervasive medium. He skillfully moved among these models as they suited him, a propensity shared by modern scientists. Apparently even Newton found it hard to drive directly to the heart of light, or he was great enough to set what foundations he could while foreseeing deep ambivalence beyond. However, when he compiled all he knew of light in his *Opticks* of 1704, he came down on the side of corpuscles, as part of a credo going back to Democritus and atomism: "God in the Beginning form'd Matter in solid, massy, hard, impenetrable, movable Particles. . . ."[10]

Newton's objection to wave theory as it was then understood seems to have been its difficulty in explaining how if light waves could bend around the edges of obstacles, they could also travel in straight lines. We now explain this behavior by a detailed consideration of wave interference, but the alternative model that relied on the straight-line motion of particles satisfied Newton as another aspect of a dynamic mechanical universe. And perhaps his views had something to do with the intrusion of human reactions into scientific objectivity. As a determined proponent of the wave theory, Robert Hooke attacked Newton's results when they were first presented, leading to forty years of enmity between the two men.

Other opposition came later from those who examined light from a different standpoint or preferred it pure rather than dismembered. Von Goethe was fascinated by Newton's experiments and tried in 1790 to reproduce them. But instead of placing a prism in a sunbeam's path, he held it to his eye and saw effects different from Newton's observations. Von Goethe framed a theory of color that sought patterns beyond Newton's mechanical understanding, such as a "law of change" to account for the tints seen in the afterimage that appears after the retina absorbs an intense picture. Blake considered Newton's objectivity to be highly limited compared to his own intense reaction to light, expressed in *Mad Song* from his *Poetical Sketches:*

> *For light doth seize my brain*
> *With frantic pain.*

In 1795 Blake made a color print showing the scientist as a naked figure drawing geometric figures with a compass. The image represents Newton's imposition of rationality on the world, to Blake an evil suppression of the imagination. Blake's labors as a professional printmaker may also have turned him against Newtonian light. Blake often duplicated works by using mechanical techniques to transcribe line and color. These he disliked, according to the Blake scholar Robert Essick, because they imposed on light and space the "graphic equivalent" of Newton's quantitative approach.[11]

Such dismissals of Newtonian thought came long after the initial euphoria over Newton's accomplishments. He was knighted by Queen Anne in 1708, the first scientist to be so rewarded, and became a national hero. Artists honored him and his work in dozens of paintings, etchings, busts, and statues. The English baroque painter Sir Godfrey Kneller painted his portrait three times. Newton appears in a stained glass window and as a statue at Trinity College, Cambridge; a second statue adorns his tomb in Westminster Abbey. Other works dramatized the prismatic analysis of light in every possible mode, from grandiose to bathetic. My favorite is the commemorative *Allegorical Tomb for Isaac Newton,* painted in 1727–1730 by the Venetian Giovanni Battista Pittoni, in which a beam of sunlight enters an interior scene full of massive arches and columns. On its way to that all-important prism, the light flows past an urn containing Newton's ashes; past weeping figures of Minerva and the Sciences, Mathematics, and Truth; past multitudes of philosophers gathered to pay homage. No grander setting could be imagined for the great experiment. At the other extreme, a work by the portraitist George Romney shows, in an apt description by the art historian Francis Haskell, "Newton experimenting with his prism while his two nieces look on with suitable expressions of astonishment."[12]

Most impressive, both in scale and in imaginative use of light, is the cenotaph for Newton proposed in 1784 by the visionary neo-classic French architect Etienne Louis Boullée. Although Boullée's most gargantuan designs were never built, they influenced the use of architecture to express national will. It was his pupil Jean Francois Therese Chalgrin who designed Napoleon's Arc de Triomphe. Boullée's plan to honor Newton placed an enormous hollow sphere, 750 feet in diameter, on a massive plinth 300 feet high; Newton's empty tomb lies at the bottom center of the sphere. From that spot, visitors see daylight through starlike openings in the sphere above, a planetarium effect simulating the cosmos that Newton had tamed. At night the sphere was to be filled with a facsimile of brilliant sunlight, although the requisite artificial illumination had yet to be invented. That enormous globe, holding a quarter of a billion cubic feet of sheer emptiness, could swallow the pyramid at Giza or a modern sports arena like the Houston Astrodome. It is only slightly smaller than the monstrous domed hall proposed by Hitler's architect Albert Speer to dominate Berlin after Nazi victory in World War II.[13]

We now see no commemorative efforts like those of Pittoni and Boullée, perhaps because science is so large and dramatic that it becomes its own event, as in a space shuttle lift-off, and its own memorial, as in the abandoned miles-long tunnel of the Superconducting Supercollider in Texas. Nor do we often see attempts to display the intangible process of scientific thinking. The varied portrayals of Newton pay homage to rational understanding of the universe and are probably the most extensive effort ever made to capture science and scientist in art. The closest modern equivalent is the widespread iconic use of Einstein's face and hair. Einstein's research, far-reaching as it is, lacks the pictorial power of Newton's prism, but the craggy, brooding statue of Einstein before the National Academy of Sciences in Washington, D.C. gives the same sense of mind contemplating cosmos as Newton's statue at Trinity Chapel. That image inspired William Wordsworth to speak of Newton "voyaging through strange seas of Thought, alone."[14]

Along with artists, the scientists and philosophers of the time applauded Newton's research in light and his other achievements. He was honored in England as permanent President of the Royal Society. His science was of the highest quality, of course, but it is also well to remember that scientists bow to received authority as well as anyone, so it is not surprising that Newton's eminence enshrined his corpuscular theory of light. After *Opticks* appeared in 1704, wave theory received little attention, whether expressed by Huygens or by the great mathematician Leonhard Euler; it did not stir again until 1801, when Thomas Young, a brilliant physician, spoke in favor of waves before the Royal Society.

Young had been a child prodigy in languages who later partially translated the Rosetta stone. His great contribution to classical light was the crucial idea of interference between waves, which he illustrated with an experiment still used to probe profound quantum riddles. Young made two small holes in an opaque barrier illuminated by sunlight streaming through an opening in a window blind. Light rays passing through each hole recombined on a distant viewing screen, making a pattern of illuminated bands like the set of peaks and valleys seen when two trains of water waves cross each other—a visible network of constructive and destructive interference. Young used interference to explain the colors of soap bubbles and revived Newton's data to find the wavelengths corresponding to the colors dispersed by the prism. But even this homage did not prevent the rejection of the new ideas as contrary to the old master's, nor did it overcome the strangeness of Young's proposal that one wave of light added to another gave darkness.

Help came thirteen years later from the French civil engineer Augustin-Jean Fresnel, who promoted waves in ignorance of Young's work and met the same cool reception. Fresnel responded by sharpening his analysis. Using the idea of interference, he scored major successes when he correctly predicted the patterns of brightness and darkness made by light bending around obstacles, and he showed in detail how radiating waves could produce straight rays of light. Fresnel acknowledged Young's priority in finding the interference principle, and the two went on to jointly explain other aspects of light by wave theory, which rapidly became accepted. By 1825, only a small minority of scientists still held to Newton's particles.

Among the successes of wave theory is its explanation of refraction—the centerpiece of Newton's investigations—as can be visualized in a gedankenexperiment. Picture a line of dune buggies racing over hard-packed sand in perfect side-by-side formation, representing the moving crest of a wave. Suppose the line of buggies encounters at an angle a patch of soft sand on which motion is difficult. The first vehicle to reach the soft area immediately slows down, as does each succeeding buggy when it crosses the boundary, whereas those still on firm sand continue at their original speed. The result is that the straight line of vehicles bends at the interface where the speed changes; similarly, each crest of a light wave bends at the interface between two media, such as air and glass, where the speed of light differs. Since light rays travel at right angles to the line of crests, the rays also abruptly change direction at the interface, becoming bent or refracted to a degree that depends on their relative speeds in the media. In air or vacuum, light of any wavelength moves at 186,000 miles per second, but the speed inside glass is lower and varies with wavelength, being greatest for red light and least for violet. Hence light is bent differently at each wavelength, the reason that Newton's glass scalpel so neatly split white light into colors.

But what were these waves that explained refraction? Although in the nineteenth century they had reached scientific favor, they were still insubstantial compared to ocean crests breaking on a beach or sound waves pressing on the ear. Light remained incomplete until work moving along another track unexpectedly converged into new insight. That came from the worlds of electricity and magnetism, which from antiquity had carried their own kinds of intangibility. Magnetism caused a lodestone to point in a northerly direction, and was thus prized by early navigators. Electricity was responsible for the invisible forces that attracted small bodies to amber that had been rubbed with fur, as the Greeks observed (amber gave its Greek name, *elektron*, to these effects). Both sets of phenomena were somewhat understood by the early nineteenth century. Electrical forces arose from the property of matter called charge, which comes in "positive" and "negative" forms. (We now know that charge is carried by electrons and other elementary particles.) The forces acted through empty space, pulling two charges together if they were of opposite type and pushing them apart if they were of the same type. Magnetism was somehow related to electricity, because an electric current—that is, a flow of charge in a wire—created a magnetic effect.

Michael Faraday, the great English experimental physicist, took the next step in understanding electricity and magnetism that unexpectedly led to comprehension of light. Born the son of a blacksmith in 1791, the self-educated Faraday's rags-to-success life makes him a storybook favorite. In 1831 he made a crucial observation that linked electricity and magnetism in a new way. He started by connecting a battery to a coil of wire to set electrical current flowing; he placed another coil of wire nearby. There was no physical connection between the coils, yet whenever Faraday connected or disconnected the battery, he detected the fleeting presence of electric current in the second coil.

Faraday knew that current flowing in the first coil created magnetism, which died away when the battery was disconnected and returned when it was reattached. That changing magnetic effect, he concluded, reached out to the remote wires of the second coil, where it made electricity flow. Faraday called this new physical principle the law of induction. This law has become central for the modern world, for it is the basis of the enormous rotating machines that generate most of our electrical power. And the discovery was essential for the Scottish theoretical physicist James Clerk Maxwell to conclude that light is a traveling electromagnetic undulation.

Maxwell was born in 1831, the same year that Faraday stated his law of induction. Bearded and serious-looking, Maxwell represented a high-water mark of mathematical physics, which he used with great power to explain the natural world. At the age of twenty-four, he reached out to far space, as he showed mathematically that the enormous rings surrounding the planet Saturn could not be solid matter, long before observation confirmed it. Those rings are the most spectacular construction in the solar system. As is true for our moon and any other planetary satellite, their behavior is set by two opposing effects: a gravitational pull toward the planet and a tendency to fly outward as they rotate. (That effect is what enables a fast-spinning laboratory centrifuge to separate the solids from the liquid in a sample of blood, for example.) Maxwell calculated that any conceivable solid ring would buckle and collapse under the combined stress from the two forces. He went on to show that only rings made of concentric circles of many small satellites would remain stable, except for a slow spreading out in space, just as had been noted over two centuries of telescopic viewing. Close-up photographs from the Pioneer and Voyager spacecraft fly-bys of Saturn in 1979–1981 clearly show the complex swarm of separate particles that makes up the rings.

This calculation shows the mysterious power of mathematics to describe reality, and Maxwell's ability to employ it. He had other scientific talents as well. Like Newton, Maxwell was drawn by color, and like Newton, he extended his boyhood love of tinkering to experiment with it. In 1861 Maxwell made the first color photograph, showing a tartan ribbon in bright shades. This interest in color became part of his greatest achievement that codified a whole class of phenomena, including light. In 1864 Maxwell published the equations he had derived to describe everything then known about electricity and magnetism. In a surprising result of stunning importance, the equations also define everything one can know about light, except for the photon to come later.

Maxwell's four equations are brief statements, each occupying only a single line, but they are written in a compressed and arcane mathematical language reminiscent of Egyptian hieroglyphics or the magical symbols that once graced an alchemist's robe. Translated into English, the first equation expresses the fact that charges produce forces like those that come from amber or make brushed hair stand up in dry weather. The second sketches the look of magnetism, as in the swirl of concentric curves that forms when iron filings are shaken onto a piece of paper above a magnet. The remaining two unite electricity and magnetism into a new compound called electromagnetism. One encapsulates Faraday's law of induction, showing how to calculate the electrical effects caused by changing magnetic effects. The other describes how electric current in a wire makes magnetism, adding Maxwell's own discovery that changing electrical conditions produce magnetism as well.

It is remarkable enough that Maxwell's esoteric marks on paper capture electromagnetism, but what do they have to do with the glory of light? The answer comes from the significance that physicists attach to the mathematical description of nature. Physics is an experimental science, yet the mathematical expression of its ideas is critical. This symbolic summary of reality makes it possible to manipulate surrogate physical properties in the mind, on paper, and with computers. Eventually the symbolic outcomes must be tested in the true world of matter and energy, but mathematics is the conceptual crowbar that pries out new relations. And just as in ordinary language, the power of analogy—the ability to use existing ideas embodied in a recorded vocabulary—can be significant in mathematical analysis.

It was the force of analogy that linked light to electromagnetism. Maxwell realized that his mathematics resembled another equation describing a well-known and tangible situation, the behavior of a wave on a string. Pluck a string stretched between two supports, and a wave crest results that travels along the cord until it rebounds at its end. The wave is transverse, which means its peaks and valleys stand out at right angles to its direction of travel. Those moving excursions set the string vibrating and give musical tones, as on a cello or a guitar. Vibrating strings have been studied since the Pythagoreans considered musical harmony in the fifth century B.C.E. In the eighteenth century A.D., scientists derived the equation that describes a wave riding on a string, which shows how to calculate the speed of the wave from the properties of the cord.

That Maxwell's mathematics resemble the equation for waves on a string suggests the existence of similar waves of electricity and magnetism. In 1873 Maxwell noted that his equations predicted a definite speed for these new waves, just as the equation for a vibrating string tells how fast a disturbance travels along it. The prediction exactly matched the speed of light, which by then had been accurately measured. That was enough for Maxwell to postulate that light consists of electromagnetic waves. And in 1888 Heinrich Hertz showed experimentally that electromagnetic vibrations in the form of radio waves moved at the speed of light. The triumph of Maxwell's equations was total. They correctly described everything then known about electricity and magnetism and gave the most developed and articulated understanding of light: It is an electromagnetic wave with undulating electric and magnetic components that travel together through space, or any medium, at a speed set by the electromagnetic properties of that medium.

Maxwell's mathematics was seminal, but physical understanding of light waves is also essential. A light wave is born whenever an electric charge gains or loses speed, as shown by a vivid image created by Faraday; he had never learned mathematical physics, but instead, he expressed electrical effects in pictorial form.[15] Faraday focused on the forces between charges by noting that any charge creates a region of space in which any other charge feels a push or pull. He portrayed this so-called electric field by drawing lines along the directions of the forces. The lines of force from a single charge radiate like the spokes of a wheel, because it would directly repel or attract any other charge of the same or opposite type. For a collection of charges, the electric field is an intricate network of connecting lines. Since charge is found wherever matter exists, Faraday's vision becomes a universe threaded through with a cosmic web of electric field lines (and magnetic field lines as well, because a similar picture can be constructed for magnetism).

Faraday made this static image into a dynamic one by taking the lines of force as elastic bands that link the charges. In this extended pictorial metaphor, if a charge jerks into motion, each line connecting it to another charge acts like a taut string that has been plucked or a whip that has been cracked: It develops a transverse kink, a changing electric field that travels along the line of force. A magnetic whip cracks as well, because the moving charge amounts to an electric current that creates magnetic lines of force with their own kinks. These linked electric and magnetic disturbances are an electromagnetic wave, moving along the lines of force at 186,000 miles per second, much faster than any speed the initiating charge could reach.

A charge makes an electromagnetic kink whenever it alters its speed, as happens when an atom vibrates in a hot body. Like a hand that continually shakes one end of a rope up and down, an oscillating charge periodically plucks its electromagnetic lines of force to make a sequence of crests and troughs. This is a complete light wave, whose frequency is the vibrational rate of the initiating charge and whose wavelength can be found from the fact that wavelength multiplied by frequency is speed—in this case, the speed of light. If the wavelength lies between 400 and 750 nanometers, the light is visible. When an eye sees light from a distant star, it senses the plucking of electromagnetic strands that begins as charges vibrate at the high temperatures within the star and ends by affecting other charges in the retina.

Maxwell's mathematics and Faraday's fields left no doubt that light was an electromagnetic wave, but they did not explain how it moved through empty space. Other questions about light as wave were to arise from the riddles called the ultraviolet catastrophe and the photoelectric effect. How these led to the photon forms a later chapter in this story, but the puzzle about the medium that carried light was the first great difficulty of electromagnetic wave theory. The problem was to see exactly how the energy embodied in light traveled from point to point. That process is easily understood in a string: When it is pulled to one side and released, the added energy is passed on to the next vibrating bit of string, and the next, like a baton handed off in a relay race. If light is a wave like other waves, its energy should also be passed on through a physical medium. But light disturbed no apparent string, agitated no obvious sea. In fact, it moved faster in vacuum than in solid glass. What, then, sloshed up and down in the emptiness as sunlight carried energy to the earth?

The properties of this supposed medium, called the ether, could be surmised before it was found. Like any material, it could be modeled as a collection of connected atoms. In hard, resilient stuff like steel, the atoms are tightly linked to one another. Energy moves quickly from atom to atom, so that any waves generated in the medium oscillate rapidly and travel at high speed. By contrast, waves move slowly among the loosely joined atoms of a soft gelatinous substance. This may appear contrary to common sense, but the experimental facts are clear: Sound waves move faster in steel than in air, in direct relation to the rigidity of the medium. Waves travel a string more quickly when it is pulled tight than when it is slack. It seemed undeniable that only a taut, resilient ether could sustain light's high speed and enormous vibrational rate. But surely such a medium would be tangible as it filled all space? And even if it could not be sensed, how could planets and stars force their way through it? Somehow the ether was firmer and more elastic than steel, yet completely penetrable.

It is probably an innate human trait to attempt repairs on an old idea, even a paradoxical one, before daring to embark on a new. Physicists expended great ingenuity in seeking an ether with the right characteristics, even imagining new forms of matter with the proper contradictory qualities. But in 1887 Albert Michelson, America's second Nobel Laureate (his Physics Prize in 1907 followed Theodore Roosevelt's Peace Prize in 1906), made a definitive measurement that led to the demise of the ether. I was reminded of his great experiment not long ago, when I visited the United States Naval Academy at Annapolis. I knew that Michelson had graduated from the Academy in 1873, but that was not on my mind as I walked among uniformed men and women, amid mementos of battles, until I came on an unmilitary surprise. A long bronze streak set near the Severn River marked the path of the light whose speed Michelson had measured when he taught physics at the Academy. Michelson was no career naval officer, resigning his commission in 1881 in favor of research, but sea duty seems to have deepened his understanding of motion. As he watched his ship move through the ocean and noted in the log the ship's speed and the prevailing currents of wind and water, he contemplated the meaning of velocity. And perhaps, while he stood watches and looked at the stars as sailors do, he thought about their light.[16]

These elements came together in the critical experiment that Michelson carried out to determine how the Earth's motion through the ether affected the speed of light. His approach can be visualized as if he commanded a ship at sea. Without fixed markers, how can the captain determine the speed of the ship in the water? He could drop overboard a line with a float on one end, as they did in the days of sail, and measure the length that streamed out in a given time. Or he could observe waves in the water, the speed of which—ignoring the effects of wind—is set by the properties of the supporting liquid. He knows that water waves always move at, say, ten miles per hour. If he measures an apparent wave speed of twelve miles per hour from his deck, there is only one reasonable conclusion: The ship is moving toward the source of the waves at two miles per hour relative to the ocean. Or if the apparent speed is six miles per hour, again only one interpretation emerges: The ship is making a speed of four miles per hour in the direction the waves move.

If the sea is calm, the captain can heave objects into the water to make waves. Once launched from the splash, the waves move at ten miles per hour, and he can again find his motion relative to the ocean. This is the experiment that Michelson carried out with his colleague Edward Morley as the Earth swam the ethereal sea. Instead of tossing weights overboard, they turned on a light source to set vibrations ringing in the ether. Instead of water waves at ten miles per hour, they observed light waves at 186,000 miles per second. But it was not easy to measure the speed of these waves as seen from the moving Earth. Our planet rounds the sun at about twenty miles per second, only one ten-thousandth of light speed. Therefore the largest effects Michelson and Morley could hope to find, if the Earth sped in the direction of the waves or directly opposite, were apparent light speeds of about 185,980 miles per second or 186,020 miles per second, respectively.

To measure these fine differences, Michelson used the interference of light waves, which is exquisitely sensitive to small deviations in speed. In the device he invented, the interferometer, light of a single wavelength is split into two beams. These are sent down separate paths of the same length and then made to reunite. At the instant of splitting, the two sets of crests and troughs coincide, but how they recombine depends on their individual journeys. If they move at the same speed, their hills and dales remain in exact step all along their twin paths. When they rejoin, they interfere constructively to give a bright spot. If the beams have different speeds, however, the peaks do not arrive in step. In the extreme, if a peak arrives simultaneously with a trough, destructive interference occurs to give darkness. The method is precise enough to discern differences in speed of one mile an hour, one-twentieth as large as expected from the Earth's motion through the ether. (This exceptional accuracy makes the interferometer preeminent in measuring minute distances and separating one wavelength of light from another. I use one daily in my laboratory.)

Michelson began to track the bright and dark spots of interference in 1881 and completed the work with Morley at the Case School of Applied Science in Cleveland in 1887. They compared the speed of light along the direction of the Earth's orbital motion, where the change they sought would be greatest, to the speed of light at right angles to that direction, where the change would be least. They made repeated measurements, allowing for every possible permutation of the Earth's movement through space (which is set by the motion of the solar system as well as the Earth's orbital speed), but they never saw the slightest variation in the speed of light. Neither have the many other scientists who later repeated the experiment. There is only one possible explanation for these results: The Earth does not move through the ether.

Yet centuries of observation and hard-fought intellectual battles tell us that the Earth does move. If it seems not to swim in the ether, then there is no such medium, or it has no meaning. The Michelson-Morley experiment and other clues indicated that there was less to the ether than met the eye. The ether did not immediately vanish, however. Along with most other scientists, Michelson did not at first see that his experiment pointed to a radical idea; rather he considered it a negative result, the chief value of which was that it inspired the invention of the interferometer.[17] In 1894, he approvingly quoted Lord Kelvin, the seventy-year-old explorer of absolute zero, who said that all the important physical discoveries had been made and that future progress would merely extend measurements to the sixth decimal place. Fortunately others came along, especially Albert Einstein, to save science from this dull fate. Within a year of Michelson's quote, the teenaged Einstein would in his mind begin to turn over the meaning of light; a decade later, he would rearrange the cosmos with his new vision of light. And the loss of the ether did slowly percolate through science. In 1900, Lord Kelvin himself declared that the Michelson-Morley experiment seriously shook his belief in the concept.

But Einstein could not have been predicted, and the tremors were not yet earthquakes. Boullée's cenotaph for Isaac Newton, proposed near the end of the eighteenth century, could be taken a hundred years later as a telling metaphor for what was then known about light. An enormous edifice of knowledge had been built. In that huge dome, the puzzles of the missing ether, of the ultraviolet catastrophe and the photoelectric effect to come, would at first seem only loose bricks in need of minor repair. The holes that Boullée pierced to bring the cosmos into his sphere also represent new illuminations that showed those apparently small flaws as large cracks. And the image of a great light-filled dome is more than a mere remembrance, because the idea of quantum light comes from profound scientific reasoning about crucibles filled with light. If Newton's spirit were to hover within the cenotaph, it might be pleased that its very memorial also symbolized the birth of photons, the latest version of his corpuscles, which along with relativity represent the modern light of the twentieth century.

4

Modern Light

For the rest of my life I will reflect on what light is.

—Albert Einstein, c. 1917

By the end of the nineteenth century, scientists could rightfully feel that the understanding of light had come a long way since the thoughts advanced by the Greek philosophers. Electromagnetic theory described every known aspect of light. The waves of electricity and magnetism replaced earlier models that alternated between particles and undulations, as Grimaldi compared light to ripples in a streaming liquid, Newton's corpuscles hurried in their straight lines, and Young's interference experiment showed that light was made of waves without specifying their nature. Maxwell's theory resolved what these approaches had not and seemed finally to give a plain physical understanding of light; yet plain understanding failed when the Michelson-Morley result ended the straightforward model of electromagnetic light vibrating in an ether.

That failure, along with others to come, could not be reconciled with the existing understanding of the universe. It took two extreme realignments to change the nineteenth-century scientific view of light: The Theory of Relativity gave the speed of light a unique property—

absolute constancy—that altered our notions of time and space; the introduction of the photon, the quantum particle of light, led to the revelation that light could be particle and wave at the same time, in some unimaginable way. These visions transformed classical light into modern light, and classical physics into modern physics, including a new quantum physics of matter.

Both revolutions in light came in 1905, near the start of the new century, and began in the same place—Albert Einstein's mind. The beginnings and ends of eras can evoke new art, but could they also provoke a new science of light? By the early twentieth century, artists were interpreting reality in nonrepresentational ways, as in Pablo Picasso's *Portrait of Ambroise Vollard* (1910), which shows front, back, and sides of the subject all at once. The art historian Linda Dalrymple Henderson notes that the angled facets of that painting are like images of a spatial fourth dimension, adding that cubist art became "more conceptual and further removed from immediate visual perception."[1] The same can be said for the science of light. Relativity and quantum physics brought it to regions where ordinary experience was at a loss, for humans had never traveled at light speed, had never encountered entities both particle and wave. Abstraction was replacing direct representation as truly in Einstein's theories as in Picasso's triangular facets. The new abstractions could even seem related. Relativity engendered the idea of a four-dimensional spacetime—three dimensions of space and one of time—which became erroneously associated with the apparent multidimensional nature of works like *Ambroise Vollard*. Einstein himself rejected the notion that relativity and cubism were linked, saying "This new artistic 'language' has nothing in common with the Theory of Relativity."[2]

Remarkably, however, a kind of direct vision did underlie the transition from classical to modern light, for the theory of relativity began as a picture in Einstein's mind. Even the most abstract science contains information about the universe first taken in through the senses, distilled though it later becomes. Newton's mechanical principles are the final concise expression of the visceral understanding that early humans gained by swinging clubs. The new interpretation of light had to wait for Einstein's exploration in his mind's eye to evolve into predictive theory.

As a child, Einstein may have had the deft hands of a laboratory scientist, for he loved to build intricate houses of cards, but unlike Isaac Newton, he did not tinker much with mechanical things. Instead, inspired by an uncle with mathematical training, he began studying algebra and geometry when he was eight years old, and calculus when he was twelve. He subsequently wrote of the "indescribable impression" that the theorems of geometry made on him, and his wonder at the realization that "pure thinking" might illuminate the real world. He pursued his own illumination when at age sixteen he envisioned himself riding a beam of light swinging through space while looking over to another beam that was traveling alongside at the same speed. As Einstein gazed from one beam to the other, he saw a paradox that became a key to the Special Theory of Relativity he would present ten years later.[3]

That paradox seemed to overturn the behavior we expect from an ordinary object, such as a train traveling along a straight track. If we say the train travels at sixty miles per hour, we mean its speed as measured by an observer standing at trackside. A passenger seated on the train interprets things differently. Gazing from her window, she sees the world vanish backward at sixty miles per hour, while she sits at rest relative to the train. And if she were to see a train moving along an adjacent track in the same direction at the same speed, that too would seem at rest, although a trackside viewer sees both trains roar past at sixty miles per hour.

Everyday experience of trains and automobiles gives us an intuitive sense that velocity is relative; that is, it depends on the speed of the observer who measures it. But what Einstein pictured violated this basic understanding. If light behaved like a fast train, then as he rode his personal light wave and looked over to the next one moving at the same speed, it should appear stationary. Maxwell's equations, however, insist that light is a set of crests and troughs that must move at 186,000 miles per second. The equations do not allow light to halt yet continue to undulate. If light were to remain light, thought Einstein, it must always travel at 186,000 miles per second, even as seen by an observer moving at the same speed (or at any speed, in any direction); to the confusion of our ingrained visceral knowledge, a rider on a light beam can never catch an adjacent beam, although he too moves at the speed of light. Einstein made this odd result an anchor of his Special Theory of Relativity, published in 1905.

Why did Einstein put such stock in his inner vision? His understanding of his own mental processes gives an answer and illuminates differences between the pre-scientific study of light and the modern approach. Einstein has written that thinking began for him when sense impressions provoked sequences of "memory pictures." A picture that appeared many times became an organizing concept that he used to understand the sensory information. And, added Einstein, these initial pictures need not be linked to words, which serve only to communicate the idea to others.[4] Aristotle and the other Greek thinkers also had their vivid pictures of light, which they described as best they could in ordinary language. But they lacked the machinery of modern mathematics, which Einstein employed to develop his imagery, and lacked the scientific belief that pictures in the mind must be verified by experiment, even if their validity seems beyond question. Early speculations about the speed of light could only be resolved by measurement. And no matter how compelling Einstein's imagery, it would carry little weight until experiment confirmed it.

The Special Theory of Relativity makes the fixed speed of light, as well as a reconsideration of the ether that supposedly carried light, into new absolutes. It was the idea of the ether that led to the expectation that the speed of light would change as seen from the moving Earth. Now the "negative result" of Michelson and Morley, the unchanging speed they measured as if the Earth did not move, had new meaning. There are different reports about whether that particular experiment influenced Einstein, who may have followed other clues.[5] In any case,

Einstein made the absence of the ether a postulate of his theory, stating that there is no space-filling medium against which to measure the speed of light. And he took as a second postulate the peculiar fact that the speed of light is measured to be the same by any observer.

Einstein displayed rare intellectual vigor in embracing these strange results just as they stood. They led him to question our deepest intuitions about space, time, and motion. Those implicit assumptions have grown over millennia but are limited to velocities far below light speed. To reform them, Einstein translated his postulates into equations and relentlessly followed their mathematical logic. The results show that space and time are not fixed, as had long been thought; rather, distances become smaller, and time intervals longer, in a moving body. The changes are minute at ordinary speeds but become drastic for speeds approaching that of light. These "violations" of experience are less extreme than they might be, for they are apparent only when we examine one moving system from another. The crew of a fast rocket ship never sees anything within the ship that offends intuition—only if they look through a viewport at another body traveling at a different speed.

This comparison between two systems moving at different speeds is central to relativity. There is a well-known photograph that illustrates the process, Jacques Henri Lartigue's picture of a fast-moving car in a 1911 Grand Prix race.[6] Through an effect created by the panning of the camera and the motion of the camera shutter, the image shows the car in a ramshackle lean to the right that distorts its wheels into ovals, while standing spectators lean perilously to the left. If the driver were to snap his own picture of Lartigue and the spectators, it would show them leaning as well. Nothing is actually tilted, of course, and the car rolls on round tires, but we know this only from prior knowledge. If an exchange of photographs were the sole communication between the photographer and the driver, each would find the other's world strangely distorted, though his own would appear perfectly normal. The exchange is a metaphor for measurements made from one system to examine another, which is how we see the relativistic changes in space and time. Intuition rejects them, but they are as real as measurement can make them. In physics, there is no deeper reality.

The relativistic effects of length contraction and time dilation have long been confirmed by experiment. Length contraction means this: If you sit beside a road and measure the length of a truck as it speeds by, you obtain a value smaller than what the driver's helper finds as he lays a measuring tape along the moving vehicle. Any moving body contracts along its direction of motion as measured by an outside observer. The body becomes shorter as its speed grows, until it reaches zero length at light speed. But the decrease is not apparent on the truck, because every part of the moving system shortens to the same degree. When the helper lays his shortened tape alongside a truck that has contracted in the same proportion, he measures the same length the truck had at rest. As for the degree of contraction, it is too small to see at any speeds that humans can attain. Even at 25,000 miles per second, far faster than any speed we can actually reach, a foot-long ruler shrinks a mere tenth of an inch. Near the speed of light, however, the effect is substantial.

Time also alters as the truck speeds past you on that highway. Measure the interval between the blinking of one digit and the next on the driver's watch, and you find that more time has passed there than on your own timepiece. To an outside observer, temporal processes run more slowly as speed increases, until they stop at the speed of light. Perhaps because time is intangible, this effect seems less believable than length contraction. It carries an emotional charge as well, for the moving clock could be a beating human heart as well as a digital watch. The twin paradox, in which one twin remains on Earth while the other travels at great speed and presumably returns younger than his brother, endlessly fascinates us. Although the meaning of this particular paradox is controversial, time dilation has been confirmed, even at earthly speeds. In 1972, an ultra-accurate clock was carried around the world by commercial airliners flying at a millionth the speed of light. In two days of travel, the clock showed the predicted changes of fractions of a microsecond (which also included effects treated in Einstein's General Relativity) relative to time elapsed on the Earth below.[7]

Much larger changes in time and distance have been measured for elementary particles, which can attain velocities near the speed of light. Energetic cosmic rays that constantly bombard the Earth from space create the particles called muons high in the earth's atmosphere, where they travel at 99.8 percent of light speed. These short-term particles typically exist for two microseconds, so their numbers decrease as they descend earthward. Only a fraction of the number measured at the 30,000-foot height of Mount Everest remains at sea level. The fraction is found to be much larger than expected, given that the muons seemingly need 31 microseconds—15 lifetimes—to cover 30,000 feet. To the muon, the Earth approaches at 99.8 percent of light speed. Applying the principle of length contraction, 30,000 feet compresses to 1,900 feet, which the muon covers in just two microseconds. More particles survive than classical physics predicts, a result that can also be understood via time dilation. To earthly observers, time on a moving muon stretches out by a factor of 16, giving a lifetime of 32 microseconds; again, more particles survive than classical theory predicts.

Now extend these effects to the full speed of light. No matter what great distance we measure for any voyage of light, to itself it covers no distance at all. The 30,000 feet from Everest's peak to sea level, the 3,000 trillion miles from the red star Betelgeuse to Earth, the twelve inches that light covers in a nanosecond—all are one and the same to a traveler on a light beam or a photon, who sees the universe approach at the speed of light. At that critical speed, all lengths contract to zero, and the traveler sees an infinitely thin cosmos. There is a corresponding strangeness about photonic time, which is infinitely dilated at the speed of light; that is, light's clock does not move from our viewpoint. Photons are timeless, existing forever in the present moment of *now,* as if speed were the preservative that keeps them immortal.

The speed of light is not infinite, as Descartes thought, but still he saw an inner truth, for infinite velocity would mean that light carries the present moment as it simultaneously occupies every point on its path. This is what length compression to zero and time dilation to eternity also imply. Descartes was wrong in scientific fact, but he was right in ascribing ultimate properties to light. To the best understanding we can muster some three hundred years later, the universe is made so that light always travels its own distance of zero, while to us its clock is stopped and its speed is absolutely fixed. These sober conclusions read as if they come out of some fevered fantasy. Light, indeed, is different from anything else we know.

Starting from length contraction and time dilation, Einstein developed other relativistic outcomes, equally difficult to grasp and accept, yet also confirmed by experiment. One result is that matter and energy are not independent but rather can be changed into each other. Einstein's equation $E = mc^2$ shows numerically that a given amount of matter (represented by m) can turn into a quantity of energy (E) that has been amplified by the speed of light squared (c^2), an extremely large number. Small amounts of mass become enormous energies, as seen in atomic and hydrogen bombs and in the radiant light of the sun that comes from the continuing conversion of its substance. Einstein also showed that the mass of a body as measured by an outside observer increases with its velocity and becomes infinite at light speed. It would require infinite energy to bring the ever-growing mass to the full speed of light—an unattainable limit for ordinary matter. Photons, however, lack mass and so do not need infinite energy.

Light is central to Special Relativity and lies also at the core of General Relativity, the theory of gravitation Einstein stated in 1916. Special Relativity is limited to motion at steady speed; General Relativity treats the more complex case of velocities that change with time. That is what connects the theory to gravity, the fundamental force that pulls masses toward one another with growing speed. Because gravity holds together our solar system and all the large material structures of the universe, General Relativity is among the most sweeping of physical theories. It is also among the most complex, for Einstein expressed it in a forbidding mathematical form that is still being explored. Recent solutions of the equations point to the theoretical possibility of time travel and show that some cosmic processes are chaotic, that is, are wildly affected by minute changes in the conditions existing at the instant they begin.[8] After Einstein's Nobel Prize in 1921 made his work widely known, it was commonly said that only a few people in the world understood relativity. It was General Relativity that was meant then, and still is.

The mathematics is complex because it describes nothing less than the shape of the universe and how that is connected to gravitation and to light. In Special Relativity, Einstein showed that the distance between two points in space, and the time between two events, depends on the observer. In the face of this complication, he found he could better describe the universe through what he called the "proper" spacetime interval, which remains the same for any viewer. The proper interval combines the three spatial dimensions of length, width, and height with time. In General Relativity, the shape of the universe is expressed in terms of a four-dimensional spacetime continuum. Light is integral to the description, for a flash of light is the fastest possible signal that can relay news of a distant event; that is, the distance light covers in a given time defines how two events are linked in spacetime. And it is light that traces out the curvature in spacetime that in Einstein's view is the origin of gravity.

Newton had taken gravity as a universal force that draws together any two bodies, a model that came from his theories of mechanics. In the Newtonian universe, celestial bodies naturally move in straight lines, except when gravitational force pulls them into curved orbits. This model correctly predicts how objects fall and planets move. But it implies that changes in position are conveyed from body to body faster even than light travels, at infinite speed. To resolve the inconsistency, Einstein envisioned a gravity without the need for forces that act through space; gravitation, he said, is embedded in the cosmos, arising from distortions in spacetime occurring near enormous bodies like suns.

An analogy shows how such distortion may cause the appearance of a force that pulls planets. Imagine a large rubber sheet stretched horizontally in midair. A prankster hidden beneath the sheet pulls down a portion to form a depression, which cannot be seen by an inquisitive eye gazing straight down from above. A marble placed on the sheet rolls down the slope, but the hovering eye sees only a marble apparently pulled by an invisible force. In Einstein's cosmos, a massive sun changes the nearby curvature of spacetime, as though it were a bowling ball deforming the rubber sheet. And like a marble swinging around the depression caused by the ball, a planet naturally falls into orbit around the sun. The distortion also affects the path of a light beam, which must follow the shape of spacetime—just as an ocean liner plying the direct route between New York and Le Havre cannot travel along a straight line, which does not exist on the Earth's surface, but must follow the arc of a great circle. A light beam traces the locally distorted contours of spacetime near any large body; light is affected by gravity.

This deflection of light became the first aspect of General Relativity to be experimentally confirmed. Suppose light from a star passes near our sun on the way to an astronomer's eye. As the sun's strong gravity affects the light beam, the image of the star should appear slightly displaced, by an amount Einstein could calculate. The actual observation had to await the special circumstances of a solar eclipse, when the moon intervenes between Earth and sun to cover the solar disk. For a short time the sky is dark, and the faint image of a star can be seen even near the sun. As a matter of pure luck, an eclipse occurred soon after Einstein's theory was published, in 1919. It was clearly visible from the island of Principe off the west coast of Africa, where an expedition sent by the British Royal Society saw that starlight was deflected by nearly the exact amount that Einstein had predicted.

Other observations (including that of the highly accurate clock that circled the world by airliner) also confirm General Relativity, although it has not received the degree of validation that Special Relativity has. Light is set into the foundations of both theories, which dominate the modern view of the cosmos. The conversion of mass into energy, and energy into mass, underlies the workings of the universe, and has done so since its very beginnings. Relativity predicts how the wavelength of light changes as it is emitted from a moving body. The resulting "red shift" to longer wavelengths tells how fast galaxies are flowing outward in the expanding universe. The effect of gravity on light indicates conditions near a star or within a black hole and shows the presence of the invisible dark matter that comprises much of the cosmos.

With his theory of relativity, Einstein embedded light in the cosmic architecture; with the invention of the photon in 1905, he explained riddles of light that arose on Earth. The roots of the photon go back to 1900, and to an elusive problem—the explanation of how light shone from hot bodies. The experimental facts were clear enough: Any hot object produces electromagnetic waves over a wide range of wavelengths. The sun is active in the ultraviolet, the visible, and the infrared; a fire radiates visible light and invisible heat. Measurements showed that a body emits more light as its temperature increases, and at shorter wavelengths. A sun hotter than ours radiates more strongly and with its light shifted toward the blue; a heating coil brightens and glows red, then orange, then yellow, as its temperature climbs.

It seemed that classical physics provided all that was needed to calculate the relation among temperature, the intensity of the emitted light, and its wavelengths. The arena for the theoretical synthesis was an imaginary crucible held at a high temperature. It was filled with light that came from atoms vibrating in its walls, to make electromagnetic waves that traveled throughout the container. This model, however, gave a strange result for the radiated light. The calculation agreed with experiment at the long wavelengths of the red and the infrared but went disastrously wrong at short wavelengths. In this so-called ultraviolet catastrophe, classical theory predicted far too high an energy, which led to the absurd outcome that a hot body radiates infinite energy at extremely short wavelengths.

Just as the paradox of the ether forced new vision, the problem with the ultraviolet led to new physics. In 1900, the German physicist Max Planck found that the catastrophe was averted if he assumed that the atoms vibrated only at certain energies, exact multiples of a fundamental unit of energy proportional to the frequency of vibration. That was also the frequency of the emitted light, which made the energy larger for ultraviolet light than for red. The greater energy required from the atoms, reasoned Planck, would mean that less power was emitted at those short wavelengths, eliminating the difficulty.

The parceling out of energy that Planck proposed violated our innate belief that any physical quantity—space, time, energy—can take any of a continuum of values. Why should an atom vibrate only at specific energies, whereas other values were not allowed? What Planck suggested was as odd as it would be to discover that water came only in indivisible units of a pint each, no more, no less, so that one could dip two pints from a lake, or sixteen pints, but never half a pint, or three and a quarter pints.

Planck's assumption was the first intimation that energy might come in discrete indivisible units—quanta—rather than a smooth flow. While this founding concept of quantum physics earned Planck a Nobel Prize in 1918, he was not pleased that the notion was at odds with classical physics, but its resolution of the ultraviolet catastrophe could not be ignored. Experiment verifies that Planck's law of radiation, derived from his packets of energy, is correct; a century later, it remains the standard way to calculate the intensity and wavelengths emitted by a hot body at a given temperature. Experiment also determined Planck's constant, the number by which the frequency of vibration must be multiplied to give the least unit of energy. This, one of the smallest constants of nature, is one of the most important, for it sets the architecture of the microscopic quantum world.

Planck's strange premise about the energies of atoms did not yet make a revolution in light. But in 1905, Einstein proposed that light itself consisted of individual quantum units later called photons, each carrying the extremely small energy given by the product of Planck's constant and the frequency of the light. The idea came from a mathematical metaphor: Einstein again considered a crucible full of raw radiation. He made a subtle calculation of the energy carried by the light and found that it resembled the mathematical expression for the energy of a gas filling a container. Einstein concluded that light was made of discrete quanta, like separate atoms of a gas rebounding throughout the cauldron.[9]

Einstein's photons were related to his rejection of the ether, for these radiant pebbles did not need a supporting medium, and they explained a deep riddle about the interaction of light with matter—the photoelectric effect. Einstein's solution was so significant that his Nobel Prize is for this work, not the theory of relativity. It had been known for some time that light shining on certain metals imparted energy to electrons within the metal, ejecting them into space. The puzzle was that these did not act as if they had been freed by waves of light. If the electrons were like stones on a beach thrown into the air by ocean waves, then with stronger electromagnetic waves—that is, brighter light—they should gain greater energy. But experiments showed that their energy depended on the color of the light, not its brightness. Blue light produced electrons of greater energy, red light less active ones. Since blue light has a higher frequency than red, this suggested Planck's association of energy with frequency.

Wave theory also failed to explain why the electrons emerged immediately after the light arrived. An incoming light wave is spread over the surface it encounters, just as an ocean wave strikes land over hundreds of feet of frontage. Each electron receives only a small part of the total energy and so would need a long time to accumulate enough to leave the metal. If an electron instantly breaks free, it is as if an ocean roller were to funnel all its immense power to a single pebble on the beach, instantly lofting it miles into the air—a marvelous, dramatic picture, and an unlikely one. Einstein replaced it with an image of incoming light as a multitude of photons, each passing its energy on to a single electron.

One might ask why we do not see the granularity of photons as we experience the impact of light on matter, which determines the colors of the world and alters rhodopsin molecules in the eye to initiate vision. The individual transactions are generally lost because each photon carries so small a cargo of energy. Even the feeble illumination from a birthday candle or darting firefly represents quintillions of quintillions of photons emitted each second. One photon in this cascade is like a grain of sand in a beachful. Examine the sand with a microscope, and you can count grains one by one; at this level, sand is quantized. Now cup some in your hand and let it run out. Separate grains are lost in the fluidlike flow, as the granularity of light is imperceptible in the flood of what we see.

Planck himself, along with other physicists, at first rejected the strange notion of packages of light. But the photon became acceptable after Einstein received the Nobel Prize in 1921 for the "discovery of the photoelectric law."[10] And it explained new results, such as those seen by the American physicist Arthur Compton in 1923. He illuminated electrons with invisible light in the form of X rays and found that the frequency of the X rays decreased after the encounter. Like one billiard ball striking another, an X-ray photon transferred some of its energy to an electron, leaving the collision with decreased energy. According to Planck's relation between energy and frequency, this corresponded to a lower frequency. Light interacted with matter according to mechanical principles; it had come full circle from Newton's corpuscles to Maxwell's waves to Einstein's photons.

Even earlier, other links had appeared between photons and matter. In 1913, the Danish physicist Niels Bohr, known for his personal charm as well as his insight, incorporated the photon into a new, quantized atom. According to the best wisdom of the time, atoms resembled solar systems, with planetlike electrons circling a nucleus. Bohr proposed that the electrons could occupy only certain orbits, as if in the true solar system no planets could exist between the tracks of Mars and Jupiter. An electron in a given orbit carried a specific energy. If an atom gained energy, that lifted an electron from an inner orbit to an outer one with a higher energy. If an outer electron dropped to an inner track, its excess energy was carried off by a photon, whose frequency was related to the energy by Planck's constant. These "quantum jumps" exactly explained all the wavelengths of light emitted by hot hydrogen gas, which classical physics had not been able to do.

Bohr's early quantum theory was later to be greatly modified, but its central ideas—that electrons occupy rungs on a ladder of energy and jump among them as photons are absorbed and emitted—still explain how light and matter interact.

The photoelectric effect, the Bohr atom, the Compton effect—these made the photon real. But there remained a problem deep and stubborn, unavoidable and inexplicable, the basic paradox: How could particles of light display the inherent wave property of interference? How could they cancel each other to give darkness? Nothing in Einstein's hypothesis explained this; in fact, as he wrote a friend, he felt his struggle with the enigma might drive him to the madhouse.[11] Characteristically, he explored the paradox in a gedankenexperiment, imagining an extremely weak light source that emits one photon at a time. In a variant of his formulation, imagine that the photon encounters an opaque barrier with two holes in it, the same arrangement that Thomas Young used to demonstrate wave interference in 1801. To ensure that only a single photon is involved, place just past each hole a light detector that clicks each time it absorbs a photon. Each click of one sensor, and not the other, guarantees that the light consists of a single particle that has traveled through that hole and not the other.[12]

Now replace the two detectors with a single strip of photographic film beyond the two holes. As each photon reaches the film, it exposes a tiny region. Wait long enough, and a pattern emerges: the same series of light and dark bands that Young saw and that was explained by interference between waves. Somehow, each single photon travels through both holes to interfere with itself. Stranger still, it seems that the appearance of wave or particle is determined by the experimental arrangement. Two detectors set to see a single photon indeed register particles of light; a strip of photographic film arranged to see interference effects shows that wavelike property.

This bizarre behavior is confirmed by experiments, such as one carried out in 1909 at Cambridge University by Geoffrey Taylor, a young physics student. He measured interference effects with light so weak that it took three months to expose a photographic plate. At most, only a few photons at a time would have participated; yet when Taylor finally examined the film, he found an ordinary wavelike interference pattern with no evidence of "graininess" in the light.[13] A definitive test requires the production and verification of a single photon, which has been achievable only in the age of lasers. In 1986 a French research team reported a modern version of Taylor's experiment, which showed that a lone photon indeed interferes with itself.[14] Light is a particle and also a wave—depending on the experiment.

That ambiguity was difficult enough to accept, but worse was to come. The wave-particle duality of light is only half a larger enigma whose other half is the particle-wave duality of matter. This stunning insight came from the French physicist Louis de Broglie, who received a Nobel Prize in 1929 for the idea. He was influenced by his older brother Maurice, the Duc de Broglie, who at his private laboratory in Paris studied how X rays affected metals. Like the photoelectric effect with visible light, the data showed that electrons were ejected by particles of light, not by waves. Louis de Broglie brought to this result new thinking he had developed in an unlikely place, a basement beneath the Eiffel Tower. Working there as a radio operator during World War I, he had contemplated the meaning of waves as embodied in radio signals.

After the war, De Broglie began to synthesize wave and particle by pointing out that even an ordinary water wave is granular at its heart, for it represents the coordinated motion of a horde of molecules. Similarly, he thought, a complete theory of light should recognize that its waves describe the collective motion of individual quanta. In 1923 De Broglie took the synthesis further when he suggested that matter was also wavelike. The idea came as he considered how to describe any type of particle, whether of light or matter. In relativity, the mass of a particle is linked to its energy; for a photon, energy is linked to the wavelike property of frequency. The chain of connections suggested that a particle with mass must always be associated with a wave. De Broglie believed the particle to be primary, for it can be observed, yet the coordinated behavior of many particles implies the existence of a wave. He called this "matter wave" an *onde fictive* when he presented the idea to the Parisian Academy of Science. When Einstein briefly considered the concept, he called it a *Gespensterfeld,* a "ghost field."[15]

These waves would not affect the common world, for as De Broglie calculated, their wavelengths are extremely small. Their effects would appear only for small-scale matter, as was first observed in 1927: A stream of electrons was directed at a metallic surface. Instead of reflecting from the surface along straight lines like tiny balls, the electrons rebounded in a complex spatial pattern; they had been diffracted by the minute grid formed by the atoms of the metal, exactly like a wave passing through an array of small openings. And De Broglie's waves showed why Niels Bohr was correct in surmising that an electron in an atom possessed only certain energies. Each orbit supported a matter wave of the correct wavelength to give constructive interference around the loop. Since De Broglie's theory associated energy with wavelength, this assigned a specific energy to each orbit—one rung of Bohr's ladder. The result showed that the wave nature of matter is intimately connected to its quantum nature.

By the end of the 1920s, it was clear that light and matter shared common ambiguous characteristics that related wave and particle and that were connected to quantum reality. That understanding grew into the modern quantum theories of matter and of light; each eminently successful, each still haunted by duality. Matter is described by the celebrated equation the Austrian physicist Erwin Schrödinger invented in 1926 to extend De Broglie's ideas. Its result, called the wave function, gives the strength of the matter wave anywhere in space and time for a given physical situation, such as an electron in an atom, but the meaning of the wave function has been elusive. In 1926, Max Born of the University of Göttingen (who, like Einstein, left Germany when the Nazis came to power) gave the interpretation we still use. The wave function is a map of potentiality, not a picture of reality. It gives the probability that an electron resides at this location or possesses that energy. The electron in a hydrogen atom is most likely to lie one-twentieth of a nanometer from the nucleus, but its wave function also spreads into other locations where the electron may also exist. The actual location of the electron, however, is found only by measurement.

A wave function is like a list of the playing cards in a deck. The list yields statistical probabilities—for instance, that over many drawings the ace of spades appears one fifty-second of the time—but cannot predict any single result. Only the concrete act of drawing a card converts potential choice into the hard fact of the ace of spades in your hand. Likewise, in what is called the "collapsing" of the wave function, a physical measurement selects a value inherent in the wave function, to give an actual result. The analogy is inexact, because drawings are random, whereas experiments measure the effects of physical causes. Nevertheless, the best meaning we have been able to give the wave function is that it represents statistical probabilities. This interpretation simply restates the wave-particle duality; it implies that the act of measurement somehow focuses a wave, which carries all physical possibilities, into a particle with definite properties.

The same statistical interpretation appears in the theory of light as photons, called quantum electrodynamics (QED). The history of QED echoes the ancient difficulties in treating light as compared to matter, for the quantum theory of light must incorporate the theory of relativity, a great complication. The roots of QED go back to the 1920s, when the Schrödinger equation was already successfully describing matter; further development of QED, however, was blocked by a mathematical difficulty, in which a seemingly reasonable calculation gave a meaningless result. In a tour-de-force of mathematical physics, the singular behavior was bypassed and the theory saved in 1948 by Richard Feynman, Julian Schwinger, and Sin-itiro Tomonaga, a trio whose accomplishment earned them the 1965 Nobel Prize in Physics.

QED is a hugely successful theory. It predicts how light behaves in space and interacts with matter. It gives a quantum view of electromagnetism, which we now interpret as arising from the interchange of photons among electrical charges. The theory also shows—in accordance with relativity—that the energy of photons can turn into particles of matter, a crucial step in the early growth of the universe after the Big Bang. Yet QED cannot escape ambiguity. Although it considers that light at bottom is particulate, this apparent resolution of the wave-particle duality simply replaces it with a statistical picture. Like the Schrödinger equation, QED can only calculate the probability that a photon will be found at a given location. Richard Feynman has said that the duality underlying this uncertainty is *"absolutely* impossible to explain in any classical way . . . [it] has in it the heart of quantum mechanics . . . it contains the *only* mystery."[16]

With this inner strangeness, it seems impossible that the quantum theories of light and of matter can succeed. The statistical interpretation, however, is accurate for large numbers of particles, the common case in the real world. It precisely predicts the properties of matter, the laws of optics, the behavior of lasers and computer chips. The duality does not prevent physicists from using the theories, for we are trained to calculate and apply the quantum probabilities. We are like insurance actuaries who accurately predict the average behavior of whole populations without understanding underlying causes of health and illness and without the capacity to describe any individual life. And quanta can offer a seemingly simple view of the world. I remember how the tiny billiard balls of quantum light delighted me as a young physics student; they seemed far more accessible than did the complexities of electromagnetic waves.

But a statistical approach merely replaces one ambiguity with another, and the simplicity of billiard-ball photons ignores the uncharted region in which they become waves. Duality writhes beneath our surface understanding of light with troubling implications. It means that the trajectory of a photon (or an electron) cannot be known; nor can we ever simultaneously measure its exact location and velocity, as expressed in the Uncertainty Principle that Werner Heisenberg stated in 1927. To many insightful minds, this indeterminism is a source of deep uneasiness. Einstein expressed his distrust of the statistical view when he said, "I shall never believe that God plays dice with the world."[17] The apparent dual nature of light and of matter, he wrote, or the equivalent statistical interpretation, was only a "temporary expedient" that would not yield deeper understanding.[18] In the same vein the American theoretical physicist John Wheeler, Niels Bohr's student, has called for the "most revolutionary discovery in science . . . the utterly simple idea that demands the quantum."[19]

That "simple idea" may have at its core a redefinition of how to combine wave and particle. One version of the quest comes down to a choice between two visions of duality. The accepted view is the Copenhagen interpretation, championed in that city by Niels Bohr in the 1920s and 1930s. It takes waves and particles as two different faces of reality. When we measure the properties of light or of matter, we see one aspect or the other depending on the experiment, but never both. J. J. Thomson, who discovered the electron, captured the idea perfectly in 1925: Waves and particles, he wrote, are like tigers and sharks, each "supreme in his own element but helpless in that of the other."[20]

The alternative vision of duality is a minority view. It harks back to De Broglie's belief that reality resides in localized particles accompanied by waves, like feathers drifting on a current of air. In this "guiding wave" or "realist" interpretation, when an electron or a photon is examined, both particle and wave should be seen. The wave function becomes directly observable, perhaps by carrying a small but measurable energy. Some recent work supports this idea. The physicist David Bohm, who died in 1992, had long been a proponent of the "realist" position. In 1957, at the University of London, Bohm and his graduate student Yakiv Aharonov derived a mathematical result about the behavior of electrons in magnetic fields. The result was thought to be unobservable, like the wave function itself, but the Aharanov-Bohm effect has been experimentally detected, showing that "unseeable" quantum effects may have physical reality. And Aharanov has recently proposed an experiment that would use a single neutron to observe a wave function before measurement has collapsed it.[21]

For light, the hope is that its duality will be clarified by the study of single photons, which fines down light to its purest state. In 1980 John

Wheeler proposed an especially incisive "delayed choice" single-photon experiment. Turn on the source, he says, allow the photon to encounter two holes in a screen, and then wait before installing either one detector behind each hole (the single photon test) or a strip of photographic film beyond the screen (the wave interference test). What, he asks, is the status of the light after passing through the openings but before detection has selected it as particle or wave? With considerable ingenuity, several research groups have carried out this experiment. The results show that light maintains its duality after encountering the holes and then is still seen as either particle or wave by the appropriate arrangement of detectors, chosen in midflight. That is, if we select the particle type of detection, the photon acts as if earlier it had passed through only one hole, but if we choose to detect for interference, the photon obligingly behaves as though it had passed through both. Can a decision to install one detection scheme or the other really change the history of the photon, or is the ambiguity in light so profound that we hardly know how to consider it?[22]

This deeply puzzling result shows that no mere handful of measurements will quickly resolve the wave-particle duality. And there are other interpretations of quantum ambiguity, in addition to the Copenhagen and realist views. These fascinate some scientists and disturb others. In the controversial "many universes" version, each possible outcome of a quantum measurement is said to generate its own full cosmos in which that result indeed occurs. A majority of physicists finds the idea of an endless sheaf of entirely real universes too rich to absorb. It violates the precept called Occam's razor, after the English Scholastic philosopher William of Occam, considered the greatest logician of the Middle Ages. About 1320 he expressed this principle of parsimony as "What can be done with fewer [assumptions] is done in vain with more."[23] Some researchers in cosmology and other applications of quantum physics, however, believe the idea of a multitude of worlds is mathematically valid, at least.

Another interpretation comes from the belief that the act of measurement shapes quantum reality, as in the Copenhagen view. Since measurement always involves a living mind, if only to contemplate the outcome, quantum physics hints that subject and object are less distinct than has been assumed in classical physics. Although this connection has been in the air since the early days of the quantum, it has not been much pursued, but these days I detect among physicists a whiff of the idea that mind influences the physical world. Immersed as we are in duality, perhaps one more dualism lies over the horizon: If neuroscientists believe that matter determines mind, physicists may come to believe—always confirmed by careful experiment—that the true meaning of the wave-particle is that mind determines matter.

The modern light of the twentieth century is defined by relativity and by the photon. Both aspects of light were difficult to believe at first, but the reality of experiment made physicists accept them. Now there is little uneasiness about relativity, only about the duality that accompanies the photon. That remaining double vision is clarified by a view from an artist. Georges Braque, who with Picasso founded the cubism that began when modern physics did, once wrote, "The truth exists—only fictions are invented," a guiding precept for those seeking to understand light.[24] Light is what it is. The scientific stories we invent to explain its maddening puzzles only reflect our present ignorance, while reality calmly continues its smooth and true functioning regardless of the tales we tell. And if mind and matter are truly linked, Braque's aphorism may carry a richer meaning; perhaps the truth about light and our fictions simultaneously invent each other.

If Braque were sitting across from me now, I would remind him of another aphorism he wrote: "Art is meant to upset, science reassures."[25] The search for reassurance explains why we seek to understand light through science, but it is difficult to feel much comfort from our present knowledge or to avoid the desperate thought that the more we stare at light, the less we see. What would Braque say if I, a twentieth-century physicist, were to assert that he is made of probability waves of matter linked by wave-particles of light, and that his perception of light depends on those same indefinable entities?

This vision brings little enough solace. Yet as Braque and Einstein were, you and I are composed of more than uncertain waves of unknown stuff. We create and carry fields of order through an enigmatic cosmos. If we do not yet understand light, our minds know how to assemble its wave-particles into comprehension. This is the reassurance that light brings. It is why we need the sun and why we must make our own light when the sun is gone. In that sense, Braque was right about science; the science of illumination, the human creation of light, comforts us. In the past, we made light as best we could. Now we create it as we wish. We use or emulate natural means—flames and hot suns, lightning and the cold aurora borealis—in candles, incandescent bulbs and fluorescent tubes, gases and solids aglow with laser light. The paradox of twentieth-century light is that it has become both more and less known. The same mysteries that seem beyond understanding also console by the light they shed, in that arena where humanity makes light.

5

Making Light

How can [Edison] call [the electric light bulb] a wonderful success when everyone . . . will recognize it as a conspicuous failure?

—Henry Morton, professor of physics,
Stevens Institute of Technology, 1879

The 1893 World's Columbian Exposition in Chicago displayed all that nineteenth-century western civilization considered great and beautiful. Its graceful white buildings exhibited culture and craftsmanship, technology and science. Electricity was a hero of the presentation, especially as represented by artificial light. At night, tens of thousands of electric lights illuminated the Exposition. Its Electricity Building was constructed around a Tower of Light whose eight-story shaft contained thousands of colored flashing incandescent bulbs. Other structures were outlined in chains of lightbulbs, and tinted arc lights shone on leaping fountains of water to give further spectacle.

It is difficult now to imagine the halo of wonderment that then surrounded the great natural force of electricity and its marvelous ability to make light. Contemporary descriptions show a childlike pleasure in electrical light, one that we have since lost. In one account, the Exposition becomes "a fairy scene of inexpressible splendor" when its "myriads of electric lights pierce night's sable mantle." An illuminated fountain "glows with the blood of the ruby . . . changes to emerald, fringed with all the shades of green the earth affords . . . jewels pale before these marvels of color. . . ." And beyond sheer response to electrical light, electricity still seemed an enormous unknown and even mystical elemental power as it flowed from laboratory into common use.[1]

Scientists of the time probably felt the same wonder at electrical light, tempered by their knowledge of its electromagnetic character and its pragmatic connection to electricity. It was electric current that made a bright arc leap between two carbon rods or that heated a fine filament until it glowed. The fundamental and the practical both appeared at a meeting of the International Electrical Congress at the Exposition, where Hermann von Helmholtz was the most famous scientific delegate. Physicist and physiologist, he had visited the territories of light again and again, in his seminal *Treatise on Physiological Optics* and in his contributions to electromagnetic theory. At the Congress banquet, Helmholtz chose to honor the achievement of incandescent electrical light. He stepped down from the speaker's table to shake hands with Thomas Alva Edison, who was scornfully dismissed by some as a tinkerer and was not even an official attendee of the Congress. But it was Edison's persistence that made electric light truly useful, as he developed the incandescent bulb along with the means to supply it with electrical current. His recognition by Helmholtz, representing the great European work in the theory of light, is a recognition of the more practical American scientific style.[2]

A century after that testimonial, we take artificial light as a given. We see it in a variety of forms that have developed far beyond filaments heated by electricity. Fluorescent lamps, lasers, and chemical processes now make what might be called cool electronic light, in achingly pure colors and glowing whiteness or at invisible wavelengths. Some modern light sources come in thousands to the square inch, some fill immense hangarlike structures. Intensities range from the imperceptible flicker of a single photon to ravening beams that cut steel. The power to make light is the reason we can suppress, but never entirely forget, the strong emotions that once accompanied the fall of night. "Each evening," says Wolfgang Schivelbusch in his *Disenchanted Light,* "the medieval community prepared itself for dark like a ship's crew preparing to face a gathering storm. At sunset, people began a retreat indoors, locking and bolting everything behind them." Darkness stirred fears of unknown forces and of human evil. In those days, to be abroad without an identifying light was de facto evidence of criminal intent. The earliest urban lighting was too dim to truly illuminate but was still needed to mark the geography of night, as it showed the presence of houses and people.[3]

This long history is the reason that light glowing from a filament heated by electricity seemed miraculous in 1893, compared to the light that had for millennia come from open flames. The Paleolithic artists of Lascaux illuminated their dark caves with flickering reddish light from lamps that have been found these thousands of years later. Most are simple pieces of limestone whose natural or reworked concavities contained the burning fuel. One has been fully shaped from red sandstone, polished and decorated with designs of interlocking chevrons. Surely it stands near the beginning of a persistent tradition that makes light sources both utilitarian and beautiful.[4]

These lamps burned animal fats such as the hard whitish form called tallow, derived from calves or oxen, which impregnated wicks of fungus or juniper wood. Modern reconstructions show that several such lamps lit at once would have sufficiently illuminated the cavern wall for painting. Lamps that burn fat or oil extend through history, using a variety of fuels and wicks. Around 1400 B.C.E. the priests and the wealthy of Egypt honored their sun god with bronze or earthenware lamps that burned olive oil. Fifteen hundred years later, priests in the city of Alexandria displayed eternal lamps with asbestos wicks that would never burn; these same wicks have also been used in the lamps of the Vestal Virgins in Rome.[5]

More centuries passed, but illumination still came from fire trapped in candles that burned solid animal products, in lamps that burned liquid oils, and in jets of gas. All made undesirable heat, smoke, and fumes along with desirable light. It is easy to dismiss these sources as primitive, but improvements came as they drew on new scientific results. Near the time of the French Revolution, the pioneering chemist Antoine-Laurent Lavoisier discovered the role of oxygen in burning, which led to a better oil lamp and then to Edison's sealed incandescent bulb. In the nineteenth century, Sir Humphry Davy discovered the electrical burning that is the basis of arc lighting. To understand all the light sources I command in my laboratory—lasers with visible and invisible emissions, fluorescent tubes, incandescent bulbs—requires twentieth-century science: Planck's law of radiation, Einstein's photons, Bohr's quantized atom.

Some advances in lighting long preceded scientific insight. Illumination was first a by-product of cooking fires but later became a separate function, with types of wood chosen for their power to make light. Candles and oil lamps added a refinement, in separating the burning fuel from a supporting structure. The solid fuel that forms a candle becomes liquid under its own heat and travels up the wick to be burned, as does the liquid fuel in an oil lamp. Within these designs, improvements came from better fuels and wicks. Gaslight also came from fuel burned in a structure, and it introduced a new feature: It replaced the individual reservoir for each light source with fuel from a central site.

Although the fuel-burning lamp can be traced back at least to Lascaux, the origins of the candle are obscure. It was known in Roman times and perhaps earlier, but there are varied theories about its invention. The idea may have originated in Africa, where oily nuts strung on twigs were burned to provide light. The first wax candles have been credited to the Phoenicians and to the Romans' northern neighbors the Etruscans, said to have repeatedly dipped a string into melted wax to build up a thick cylinder.[6] Whatever their origin, by the Middle Ages candles had moved from limited use in religion and ritual—as in the ceremony of exorcism denoted by bell, book, and candle—to become the main source of artificial light. Then and into the nineteenth century, however, they were too expensive for regular use. And they never gave enough light to read comfortably at night, although they were improved in other ways. In 1825 Michel Eugène Chevreul, the chemist in charge of dying operations for the Gobelin tapestry works, and Joseph-Louis Gay-Lussac, who discovered an important law of gas behavior, patented a smokeless candle.[7] The great nineteenth-century scientist Michael Faraday also pursued the complexities of the candle. In a famous lecture published as *The Chemical History of a Candle,* he explained why a fluted candle is beautiful but ineffective, and he described the intricate zones of heated air in a candle's flame.[8]

Although candles never became brighter, oil lamps did. Near the end of the eighteenth century, the Swiss chemist Francois Ami Argand enhanced the light from oil lamps by applying the new theory of combustion developed by Antoine Lavoisier. In the era before the French Revolution, Lavoisier founded much of modern chemistry while serving in the quasi-governmental institution of the Farmers-General. His scientific interests merged with social concerns in his efforts to improve farm lands near the Loire Valley, in an award-winning essay about lighting urban streets, and in his contributions to the new metric system of measurement. The holder of a title and involved in tax collections through the Farmers-General, Lavoisier fared badly in the Reign of Terror. He was attacked by revolutionary extremists, including Jean-Paul Marat, and was guillotined in 1794.[9]

Lavoisier's greatest scientific achievement upset the one-hundred-year-old theory of combustion that depended on phlogiston, a hypothetical material thought to be given off in burning and to leave behind a residue. Lavoisier showed that the remaining ash was a compound of oxygen, and that oxygen was essential for burning. This inspired Argand, who had briefly studied under Lavoisier, and in 1783 Argand devised a hollow wick for oil lamps that brought an additional flow of air, which is one-fifth oxygen, to the inside of the flame. (Benjamin Franklin had attempted a similar modification but apparently with too small a wick.[10]) A second improvement was the addition of a glass cylinder as a chimney to increase airflow around the outside of the flame. A third was a means to raise and lower the wick, which changed the size of the flame and hence the amount of light.

This enhanced lamp was quickly accepted. A year after its invention, Thomas Jefferson brought home from France a version crafted in silver. The Argand lamp was valued for its brilliant light, perhaps in its time as stunning as laser light is now. Its oxygen-fed flame burned at a higher temperature than in any candle or other lamp. This produced more light, and of a whiter hue, as we now understand from Planck's law of radiation. The high temperature also consumed the smoke of carbon particles that dimmed and fouled earlier oil lamps, an important selling point in contemporary advertisements.[11] Protected from air currents behind its glass cylinder, the flame burned with remarkable steadiness. The lamp operated so well that it was used in lighthouses until the late nineteenth century.

Beyond the sheer fact of its brighter light, the new lamp had wide cultural influence. Just before the French Revolution, bright Argand lamps enhanced the great glitter and draw of the Palais Royale, the center of Parisian nightlife. The cheek-by-jowl contact there reflected tensions among social castes and eroded barriers between them.[12] The lamp also helped people to learn. L. Sprague de Camp, a historian of technology, comments that although the perfection of the printing press in the fifteenth century made books cheap and plentiful, a true revolution in reading also required eyeglasses, invented in the thirteenth or fourteenth century, and the bright Argand lamp. These gave folk new opportunities to read and learn at night after the day's work was done and to carry on lifelong reading even as age diminished their sight.[13]

With its individual store of fuel the Argand lamp lacked an element crucial for electric light to come—a distribution system. Gaslight introduced the idea of a network of lamps tapping a central reservoir of energy. Illuminating gas was first recognized as a by-product of the processing of coal to make tar. In 1806 gas lamps were installed in a cotton mill in Manchester, England. Their use spread quickly. By 1815 many miles of pipe carried gas from a central generating station to lamps in the streets and in a few private residences in London, where they illuminated by burning gas at a perforated orifice. Fifteen years later London was illuminated by thousands of gas lamps, which were seen as a great deterrent to crime.

Gaslight was much brighter than light from candles and oil lamps. Some found it even too strong, especially after the Welsbach mantle—an incandescent sheath placed around the flame that glowed with exceeding brightness—was introduced in the 1890s. Gas eliminated the bothersome tending of wicks and gave an even, unflickering light, but it had its own problems. The huge gas storage containers were always thought to be in danger of exploding. Large indoor installations, such as the thousand gas jets that illuminated one Parisian theater in 1862, raised temperatures and released fumes as they used up oxygen. Theatergoers would develop headaches in the poor atmosphere. Eventually theatrical gas lamps were built into ventilation systems, but this could not be done in the London Underground, the carriages of which were lit by gas carried in bags on their roofs (passengers also brought their own candles), further distressing an atmosphere already made noxious by the steam locomotives of the system.[14] There were other undesirable effects. In 1850 Faraday noted that burnt illuminating gas was damaging paintings at the National Gallery in London. In the Sistine Chapel, centuries of lighting by open flame dimmed the *Last Judgment* until it had to be cleaned more than once.

Electrical light finally overcame the difficulties of light from open flames. Light from electricity was first seen in the nineteenth century from the carbon arc, discovered by Davy at the Royal Institution in London. Later knighted for his varied scientific work, Davy isolated potassium and other elements, found the anesthetic properties of nitrous oxide (laughing gas), and invented a safety oil lamp for miners that reduced the danger of igniting underground gasses. His brilliant lectures inspired Faraday, whose career began as Davy's assistant. In 1801 Davy attached two carbon electrodes to a massive electrical battery and saw a spark jump between them when he held them a fraction of an inch apart. The flow of current through the air continued as he further separated the rods to a distance of several inches. It heated carbon particles to incandescence, consuming the rods to make a brilliant white glow. Not long after, Davy observed that an electrical current passed through platinum heated the metal until it glowed with incandescence. Fully developed incandescent lighting, however, came only after arc lighting had been explored.

The electrical burning that makes arc lighting could be clearly seen in the version called the Jablochkov candle, after its inventor, the Russian engineer Paul Jablochkov. Two parallel carbon rods were held together with an intervening layer of insulator. They were electrically ignited at one end and indeed burned down like a candle. The light from such an arc is enormously bright. Women would open their umbrellas to protect themselves from its rays, and gaslight seemed by comparison "dim . . . red, and sooty."[15] The power of the arc was especially appreciated in Paris, called the City of Light from its long history of street lighting (ordained by decree of Louis XIV in 1667) as well as its role in the Enlightenment. Planners for the 1889 Paris Exposition heard a proposal to build near the Pont-Neuf a Sun Tower a quarter of a mile tall. Fitted with arc lights, it would illuminate all Paris.[16] Gustav Eiffel's tower was built instead, but it too was originally planned to mount great arc lights at its top. If these plans were impressive only in conception, similar schemes were actually carried out in the United States. In the 1880s the cities of New York, Cleveland, and San Jose erected arc-lighting towers up to twenty-five stories high but with disappointing results, for their illumination resembled only pale moonlight. Arc lights set on ordinary lampposts were more practical, and they lit urban streets well into this century.[17]

Arc lamps could not light whole cities, but they were still too powerful for indoor use. In the terminology of the time, it was impossible to "subdivide" arc lighting, for if the voltage was reduced, the arc ceased. Electrical light became universal only with the incandescent bulb, the last great source of light from heat, which began the modern science of illumination. Incandescent electric light could be subdivided; safely contained in a sealed globe, it did away with open flames and burning carbon. This new light was developed in 1879 through the combination of technological vision, drive for commercial success, and empirical method that was Edison's trademark.

Edison was largely self-taught, through a program of solitary reading that was probably encouraged by the deafness which began at age twelve. As a teenager, he attempted the *Principia Mathematica,* Newton's masterwork that lays out his theories of mechanics and gravity in highly analytical form, and came away with a strong dislike of mathematics; Edison was later inspired by Faraday's nonmathematical writings. However, there is more to Edison's style than the difference between laboratory scientist and theorist. He found a need and made the invention that filled it, as he did with telegraphic devices, the stock ticker, the carbon microphone. He was a technological entrepreneur, a natural outgrowth of a life in which he made his way from early on with varied work, including that of telegraph operator. His laboratories at Menlo Park and West Orange, New Jersey, pioneered modern industrial research; they were organized to solve scientific problems in the expectation of profit. The animating ideas flowed from Edison himself and showed a different insight than Newton or Faraday could have offered.[18]

Edison focused on the incandescent effect when he first considered electric light in 1878. Characteristically, he obtained venture capital by forming the Edison Electric Light Company even as he attacked the technical problems. One concern was that the glowing incandescent source, the filament, had to reach an extremely high temperature to make bright white light. According to Planck's law of radiation, a hot body shines over a range of wavelengths that become shorter at higher temperatures, while the light intensifies. An electric stove burner at 100 degrees Celsius, the temperature of boiling water, emits radiant infrared heat but no visible light. At 700 degrees the burner becomes visibly red-hot; higher temperatures introduce the shorter wavelengths of orange, green, and so on. The mixture of colors looks increasingly "white" until at 2,500 degrees (far above what an ordinary stove reaches) the glow is intensely white-hot. Early researchers found that electrically heated platinum became hot enough to give brilliant white light, but it melted at a temperature only slightly higher. If the current deviated and the temperature rose, the filament destroyed itself. Carbon had a higher melting point but could not be used because it oxidized or burned as it operated at white heat.

Another issue was the filament's resistance, the property of a material that makes heat and therefore light as electric current passes through it. Although not known in Edison's time, we now understand the underlying microscopic process—it is akin to the generation of heat by mechanical friction, as when the brakes of an automobile become hot after hard stopping. Picture electric current as it really is, a stream of electrons pushed through a wire by a voltage. The traveling electrons collide with the equivalents of barriers and potholes—the atoms composing the wire and the places where atoms are missing or are replaced by impurities. Each encounter converts energy from the moving electron into heat, or vibratory motion of the atoms. The oscillations shake Faraday's electromagnetic web to make light. The more frequent and numerous the collisions, the greater the resistance, which means more heat and light at a given flow of current.

There is an important pragmatic aspect to resistance that Edison keenly appreciated. Low resistance means high current, as a smooth road accommodates a greater stream of automobiles than does a rough one. Before Edison, an incandescent lamp was typically made with a thick filament of low resistance, which drew high current from an individual battery. Edison, however, wished to link his lamps into a network that took power from a central source. With high currents, energy would be lost as wasteful heat in the connecting wires. Edison planned to use low-current lamps with thin, high-resistance filaments. Each would operate from its own current loop, so that even when a lamp failed, the rest would continue to glow. Edison considered high-resistance lighting the truly novel feature of his system. The innovation drew on the calculations of one Francis Upton, who had been trained in physics by Von Helmholtz and brought that theoretical background to Edison's team.[19]

Edison sought a high-resistance, high-temperature filament that could emit strong white light without rapid deterioration. He tried platinum but could not prevent it from melting; other metals also proved unsuitable. He turned to carbon and in 1879 found a working filament. It was a loop of cotton thread that had been carbonized, or baked in a furnace until only a thin strand of carbon was left, a strand that provided high resistance. To prevent oxidation, the filament was placed in a glass globe that was then evacuated and sealed. The first carbon-thread lamps emitted light for forty hours before they failed. They were exceedingly fragile, and Edison used unsubtle means to search for sturdier filaments. Anything containing carbon was worth trying. This includes a whole universe of organic substances, and Edison tested wood of many sorts, cork, tissue paper, coconut shell, human hair, and more. Carbonized cardboard gave lifetimes of 170 hours, but in 1880 Edison found that carbonized bamboo was even better; it served until 1889, when other materials came on the scene.

Edison's successful lamp also required the separate technical achievements of evacuating and sealing the bulb and of designing an electrical generator, which consisted of a steam engine that rapidly rotated a coil of wire in a magnetic field. This was not the first machine to use Faraday's principle of induction to make current, but Edison's versions were the most efficient yet. The entire lighting system was tested in 1880 on the steamship *Columbia,* where it was advantageous for small enclosed compartments. For general use, Edison's bulbs were field-tested in the London Underground before his installation in lower Manhattan began operation in 1882. This complete network was modeled on the gas system—Edison had even contemplated using existing gas channels and chandeliers—and provided electrical current for more than one thousand lamps used by some fifty customers. Other uses for the new light came quickly. In 1881, a thousand incandescent bulbs made by the British inventor Joseph Swan illuminated London's Savoy Theater for a new production of Gilbert and Sullivan's *Patience.*

Modern lamps return to metal filaments. Since 1912 they have used tungsten, its high melting point allowing it to shine at white heat for hundreds of hours. In the tungsten-halogen lamp now used in automobile headlights and homes, a small amount of iodine reduces deterioration of the filament, which permits higher temperatures with whiter light. Nevertheless, there are serious limitations in incandescent light: It is inefficient, for much of the electrical energy put into a bulb emerges as radiant heat rather than visible light. Nor is it easy to control its color. But Nature supports other means we can use efficiently to make light without heat, at specified wavelengths. This cool light is said to be luminescent rather than incandescent. It could also be called electronic rather than electric, because it derives from the quantum interaction of electrons and photons.

Edison lived before the discovery of elementary particles, but in a way his light begat the age of the electron. In 1883 Edison examined a small mystery in his incandescent bulb, the appearance of a dark spot on the inside surface. He found that the hot filament emitted negatively charged particles, but he did not pursue the observation. This Edison effect, called his sole scientific discovery, is now understood to show the emission of electrons from the filament. It is the heart of the vacuum tube, the invention responsible for radio, television, radar—in fact all electronic science until the 1950s, when the transistor began to replace that fragile glass cylinder with the robust solid-state devices we now use. The same observation, made by a researcher driven by general curiosity rather than focused aim, might have accelerated the discovery of the electron. But the mind that would enjoy such free inquiry would not be the mind to develop a complete and practical lighting system.

Had Edison continued on to the subtle use of the electron, he might have given the world an early understanding of cool light. Nevertheless, that light was seen before electrons and photons were known, by the British chemist and physicist William Crookes, who made a career of the intangible. He adeptly manipulated the nothingness of vacuum, measured nearly imperceptible light energy, studied and perhaps believed in psychic phenomena. His scientific reputation came from his discovery in 1861 of the element thallium, which involved the weighing of the new material under vacuum for utmost accuracy. Vacuum appeared again in his "light-mill" of 1875, a radiation detector still seen today as a toy. With its four vanes, each black on one side and silver on the other, it whirls merrily in its evacuated glass sphere when it is illuminated.[20]

These skills led to the Crookes tube, a highly evacuated glass tube with an electrode at either end. It was crucial in the later discoveries of X rays and electrons, but it served another purpose for Crookes. In 1878, the year Edison began work on his bulb, Crookes found that if he filled the tube with gas and applied a high voltage, the electrical discharge made the gas glow with light. By 1900 gas-filled tubes were used for lighting, and in 1910 the French chemist Georges Claude showed that he could excite neon gas into a brilliant red glow. Neon light soon appeared in advertising signs and formed part of a whole spectrum of colored-discharge lighting from different gasses and vapors: blue and green light from argon gas, orange and green from krypton, blue from xenon, a warm yellow from sodium vapor.

The specific wavelengths of discharge lighting come from quantum jumps of electrons in atoms, as Bohr explained in 1913. Energy added to an atom by electrical voltage excites an electron to a higher energy level. A short time later, the electron returns to its original state. As it does, it gives up the extra energy as a photon of a specific frequency, defined by Planck's relation between energy and frequency. The nature of the particular atom determines its energy levels, so each kind of atom has a unique set of quantum jumps and emitted wavelengths, such as the red of neon.

The striking colors of discharge lighting would not do for general illumination, as people want white light that approximates sunlight. That came from a refinement of discharge lighting that uses the process of fluorescence, in which light is absorbed at one wavelength and re-emitted at others. Antoine-Henri Becquerel, who in 1903 shared a Nobel Prize with Marie and Pierre Curie for the discovery of radioactivity, made a crude fluorescent lamp in 1867; Edison also examined the method. But fluorescent light did not appear in commercial form until the Chicago Centennial Exhibit of 1933. By 1950 fluorescent tubes had partly supplanted incandescent bulbs.

In a fluorescent lamp, the light that comes from mercury vapor in a discharge tube is redefined into acceptable form. A voltage applied to mercury vapor excites quantum jumps that produce blue-green light. That is the color seen under mercury-vapor lamps used outdoors, where high intensity is more important than the bizarre black lips and gray cheeks the light imparts to faces. In the fluorescent tube, this ugly light is intercepted by a phosphor, a fluorescent coating inside the tube. The energetic ultraviolet photons raise electrons in the phosphor to higher levels. As the electrons drop back to lower energies, they emit photons whose wavelengths cover much of the visible spectrum to give white light.

Fluorescent lighting is more efficient than the incandescent form because it converts a larger fraction of electrical power into light. But the earliest phosphors gave more blue and less red than sunlight or incandescent light. People did not like their appearance under this illumination that lacked warm tints and gave strange color renditions. Fluorescent lighting is not the first new type of illumination to seem less pleasing than older forms. Nineteenth-century theaters advertised their expensive seats as illuminated by candlelight that flattered complexions, rather than by harsh gaslight, and the last London theater to change from gas waited until 1902 because its leading actress found electrical light harsh.[21] Fluorescent lamps, however, introduced something completely new—flat lighting without three-dimensional modeling. The divergent rays from a small flame or incandescent filament cast shadows and show contours. The long fluorescent tube gives even lighting, excellent for the workplace, but it casts few shadows and sets no mood. Now fluorescent light comes in more pleasing forms, using new phosphors that give warmer coloring and small tubes for fuller modeling.

The cool, flat illumination from a long fluorescent tube can shape the nocturnal world, as Edward Hopper showed when he put this new light into his *Nighthawks* of 1942. Hopper's artistic desire was to "paint sunlight on the side of a house," and he brilliantly portrayed sunlight as it lay across a wall or entered a room.[22] Many of his best works, however, show artificial light. The diner in *Nighthawks* is isolated and adrift in urban nighttime darkness. Its cargo of habitués—a couple and a lone man seated on counter stools, a counterman dressed in white—is oddly remote from the viewer. That mood comes partly from the brightly garish, shadowless fluorescent illumination inside the diner, in contrast to light from a street lamp shining on an exterior wall of warm red brick.[23] The art historian Robert Hobbs calls the fluorescent light in the scene "alienating and dehumanizing . . . clinical. . . ."[24]

Other works by Hopper show different aspects of artificial lighting. In *Summer Evening* (1947), a man and woman talk as they lean against the inside of a porch railing. Light from a round incandescent fixture seems trapped within the enclosure, separating it from the surrounding night.[25] *New York Movie* (1939), which shows a cross-section through the artificial lighting in a theater, could be a treasure to future historians of illumination.[26] At far left we see a segment of movie screen, where manipulated light creates an enormous black-and-white image. The audience is barely suggested under a dim reddish glow from overhead fixtures, but an usher is dramatically lit as she leans pensively against the wall of an adjoining corridor. Her blond hair glows in the bright light from three incandescent lamps set in a wall sconce, like electric candles. Other interior scenes by Hopper show receding hanging globes of light that echo images by Degas and Manet and show the hidden indirect lighting that was coming into vogue in the 1930s and 1940s.

It is not only Hopper's art that shows the artificial lighting of a given time and how that light influenced the look of the world at night. In 1912 and 1913, Robert and Sonia Delaunay (also known as Sonia Terk) walked the Boulevard Saint-Michel in Paris, gazing in amazement at the newly installed electric lamps that they saw as surrounded by halos. Sonia expressed their perceptions in her *Electric Prisms* of 1914. It is filled with bright rings of color around each lamp, as if a kind of Newton's prism were decomposing light. The piece reflects Robert's beliefs about the dynamic effect of simultaneously viewing several colors. These hark back to a theory of color due to Chevreul, the same expert in the chemistry of dyes who helped develop a smokeless candle.[27]

Gaslight also appears in art, even in this century. Several works by Marcel Duchamp portray gaslight or allude to it, including one of his Readymades and his final piece, the installation assembled from 1946 to 1966 and residing at the Philadelphia Museum of Art called *Etant Donnés: 1° La Chute d'eau, 2° Le gaz d'éclairage*. It shows a reclining nude set against distant wooded hills and a waterfall and clutching aloft in one hand a small glowing gas lamp. The work suggests both water and illuminating gas as flowing from connected networks.[28]

Nineteenth-century art richly recorded the look of gas lighting. In his *Le Café-concert des Ambassadeurs* (1876–1877) and other works, Edgar Degas showed chandeliers, glowing globes of gaslight, the strong upward cast of theatrical footlights shining on performers. The last was an outcome of the lack of sources that could project intense light. Only footlights at stage front could be set near enough to illuminate singers and actors. But faces seen by light that shines up look strange, because we usually read faces in light cascading down from the sun. Nineteenth-century theatrical makeup was designed to compensate for this. Electric light, however, gave better simulations of reality, a significant contribution to a naturalistic twentieth-century theater. Even without electric light, nighttime Paris could glitter under gaslight. Édouard Manet's *Bar at the Folies-Bergère* (1881–1882) shows, along with the barmaid gazing full at the viewer, a brilliant scene illuminated with a chandelier and spherical fixtures. And as some nineteenth-century artists chose to show the new world of machinery driven by steam, Paul Signac—who with Georges Seurat founded the pointillist technique of painting with dots of color—treated the machinery of gas lighting. His *Gas Tanks at Clichy* of 1886 shows huge storage cylinders looming behind a house.

Other nineteenth-century work also hints at the state of illumination and its meaning in society. Etchings in Gustave Doré's *London* series of 1872 show the crudely focused beam of a bull's-eye lantern; gaslight in the streets, at a night refuge for the indigent, and at a ball; and the works in Lambeth where illuminating gas was made.[29] Around 1820 the British illustrator George Cruikshank, with his brother Robert, made images for the book *Life in London* that present the contemporary social rankings of illumination: candles in chandeliers illuminate a ball scene populated by the ladies and gentlemen of "high society"; gaslight illuminates a "gin palace" with less elegant but still mostly presentable patrons; lowest on the scale, oil lamps serve another bar with far racier customers.[30]

Artists of the sixteenth and seventeenth centuries probably found it difficult to work under light from candles and oil lamps, but some painted superbly under that limited illumination. Many were influenced by Caravaggio, the sixteenth-century Italian famous for his intense lighting effects. It was said at the time that Caravaggio painted in a shuttered room illuminated by a single lantern.[31] True or not, many of his works, such as *The Sacrifice of Isaac,* show a beam of light as from a lone source, illuminating the central figures to heighten drama. One of many Caraviggisti who followed was the Lorraine painter Georges de La Tour, appointed official artist to Louis XIII in 1639. His painted candlelight intimately models his subjects in warm tones—the antithesis of flat fluorescent lighting—as in *The Penitent Magdalen* of c. 1640. In *St. Irene Mourning for St. Sebastien,* De La Tour shows the dynamic twisting flame of a torch with marvelous effect. His pictures record other sources of light: a burning wick afloat in a tumbler of oil, an ornate lantern containing a single large candle.[32] Lamps played a special role in the *vanitas* ("futility" in Latin) style of painting, which portrayed the transience of human existence. Harmen Steenwyck's *Still Life: An Allegory of the Vanities of Human Life* (1612–1655) contrasts the worldly pleasures of musical instruments and precious objects against signs of mortality: a skull, a chronometer, a brass lamp emitting a thin trickle of smoke as it burns the last of its oil.

In this century, new forms of illumination have turned light itself into artistic medium. The American artist Dan Flavin first used fluorescent light to make a glowing sculpture in his 1963 piece *The Diagonal,* which placed a single lamp at an angle. His later works are ornate installations of white and colored fluorescent lamps and neon tubes that delineate space and color in a room. The contemporary California artist Robert Irwin has used fluorescent light in what he called "spatial drawing;" in his *Three-Plane Triangulation* (1979) long, glowing lines of fluorescent lights form grids of illumination that guide the eye through space.[33] Other artists have moved away from the linear lamp. In his *Lit Circle* (1969–1986), Keith Sonnier combines the curved arc of a formed neon tube with a painted glass disk. Bruce Mauman's *The True Artist Helps the World by Revealing Mystic Truths* (1967) uses neon tubes shaped into a spiral and into flowing script.

James Turrell has extended the artistic use of light in a different manner. In the late 1960s, he combined artificial and natural sources to create light as a seemingly tangible, space-filling entity. His work, wrote Turrell, "is made of light. Light . . . is itself the revelation." His *Afrum* (1967) seems to be a large, solid cube of light afloat in space, yet on closer examination it becomes only light projected onto a wall. In his "shallow space" constructions, such as *Ronin* (1968), fluorescent lamps set behind partitions flood the space beyond with bands of illumination and create the illusion of free-floating walls. Like the work of Flavin and Irwin, these images require specific technology. The projected pieces are more effective with an intense xenon source than with an incandescent one; the shallow-space works use the even, linear lighting of fluorescent tubes.[34]

Expressive uses of artificial light are not limited to visual art. In versions more or less refined and powerful, they appear in the theater, *son-et-lumière* displays, rock concerts, in Times Square and Las Vegas. Artificial light attains its greatest impact when it conveys authority, pomp, or political might. Such dramatic use of light is probably as old as illumination itself, and it conveys the same messages regardless of the kind of lighting. Control of the light confers a spiritual mastery that can illuminate the interloper and the unbeliever or place them in outer darkness. Anyone who has been skewered, as I have, by the million-candlepower beam of light stabbing down from a hovering police helicopter knows that feeling of exposure and helplessness. Artificial light also shows power when it represents the profligate use of resources to achieve nothing but ephemeral radiance: In 1688, Louis XIV illuminated the park at Versailles by the flames of twenty-four thousand dearly expensive candles, a display of ostentatious waste appropriate to the Sun King.[35]

The urban illumination that Louis decreed in the City of Light also came to carry political meaning. By Revolutionary times Parisian streets were lighted with *réverbères,* oil lamps whose reflecting surfaces efficiently directed their rays. These were under police jurisdiction, and it was considered a serious crime to damage a lamp, thus the Parisian mob naturally associated the lamps with authoritarian government. The mob smashed them during the Revolution and hanged its victims from lamp standards—a form of summary justice extolled in the song "Ça Ira"; the *réverbères* became part of Revolutionary folklore. Illumination also entered into the Revolutionary penchant for enormous constructions, as it had in Boullée's huge cenotaph for Newton. In 1799 one Dondey-Dupré presented to Napoleon a plan to light all Paris: The main illumination would come from a lighting tower located at the Place de la Revolution, with secondary towers arranged throughout the city in such a way that no shadows would be cast. The proposal alluded to a secret fuel that would generate this great light, a fuel that may have been the then-new illuminating gas.[36] Gaslight did come to be used at the other end of the political spectrum, as thousands of jets illuminated St. Paul's Cathedral during the Duke of Wellington's burial in 1825.[37]

A century later, the Nazis used the power of light for their own ends. Their party rallies of the 1930s included thousands of marchers carrying torches or parading beneath flags illuminated by spotlights. The Nuremberg rally of 1937 featured 130 anti-aircraft searchlights that Hitler's architect Albert Speer borrowed from Hermann Goering's Luftwaffe. Speer arranged them around the perimeter of the meeting field with their shafts of light pointed straight up, defining a huge volume extending thousands of feet into the air. That space carved out by light, a "cathedral of light" as Speer called it, moved the British ambassador Sir Neville Henderson to deem the effect "solemn and beautiful."[38]

But now lasers, not searchlights, are the most powerful expression of artificial light, political or otherwise. Their concentrated light is strong enough to cut metal, subtle enough to carry information, pure enough to inspire scientists and artists. The laser in all its varied forms gives the most intense and versatile light we have ever made. Its name is an acronym for Light Amplification by Stimulated Emission of Radiation, the physical process that makes the laser unique among light sources.

That amplification of light can occur among the atoms and molecules of solids, liquids, or gases. It uses the same quantum leaps among energy levels that make photons in discharge lighting, where the process is called spontaneous emission. The laser extends this scenario through a second quantum effect, stimulated emission, which increases the number of photons. In 1917 Einstein noted that an electron residing in a higher energy level in an excited atom is like a boulder at the edge of a cliff that can be made to fall by the merest tap. In the atomic world, that nudge comes from a passing photon, which stimulates the downward jump of the electron. The descent produces a second photon identical to the first in wavelength, direction of travel, and phase. Four decades later, this idea led to the laser, in which stimulated emission causes a chain reaction where one photon becomes two, each of which triggers another two, and so on. The cascade grows into intense, well-defined, coherent light unlike that from incandescent and discharge processes, which do not create identical photons at a fixed and exact wavelength.

The crucial elements in the birth of the laser combined industrial and academic research. Charles Townes at Columbia University and

Arthur Schawlow at Bell Laboratories first discussed in 1958 the use of stimulated emission to make light (both were later awarded Nobel prizes, Townes in 1964 and Schawlow in 1981). In 1960 Theodore Maiman of the Hughes Aircraft Company built the first laser, using a solid rod of red ruby (the compound aluminum oxide with a trace of impurity) as the medium in which light was generated by quantum jumps. But earlier, in 1957, Gordon Gould had conceived a different kind of laser, with a gaseous medium. Gould, a graduate student in physics at Columbia University, had designed an improved type of contact lens and considered himself an inventor. He entered his idea for a laser into a laboratory notebook—where the record shows that he also invented the acronym "laser"—but he did not immediately seek a patent. After a long legal battle, he received the patent in 1987.[39]

A workable laser requires a medium with many excited atoms, so that stimulated emission can occur. But excited atoms tend to lose energy by spontaneous emission, giving many at low energy and only a few excited ones. The trick is to maintain a majority of excited atoms at all times. One method appears in the helium-neon or HeNe (pronounced hee-nee) laser, whose bright red beam is emitted from supermarket checkout scanners. The red light comes from neon gas in a tube, excited by high voltage as in a neon sign. Mixed with the neon are helium atoms that also gain energy from the voltage. They pass on the energy to neon atoms as they collide with them, creating enough excited neon atoms to sustain stimulated emission. To generate as much light as possible, the ends of the tube are capped with mirrors that endlessly reflect photons back and forth through the gas. Similar designs work well for lasers charged with other gases, such as a tube in my laboratory filled with argon gas. The gas glows under voltage, and transitions among its energy levels produce laser light in the ultraviolet, the blue, or the green range.

Laser action can also occur among the energy levels of molecules rather than atoms, as in a second laser in my laboratory, a five-foot-long tube filled with carbon dioxide, CO_2. The three atoms of the CO_2 molecule vibrate as if they were linked by tiny springs. Like any energy of the microscopic world, the vibrational energy appears as a set of distinct quantum levels that sustain stimulated emission and make laser light when the gas is energized by a voltage. The carbon and oxygen atoms vibrate relatively slowly, giving light at the low frequencies and long wavelengths of the infrared. Different quantum jumps among the levels can produce any of a hundred different infrared wavelengths, although the strong emission at 10.6 micrometers is most commonly used.

I can choose any of these wavelengths by changing the optical characteristics of my CO_2 laser. This optical tuning demonstrates quantum reality with visceral impact. I monitor the invisible infrared beam by passing it through a modern version of Newton's prism, which placed red light and blue onto different locations on his wall. My device changes infrared light into visible form and displays it on a screen at a location that depends on wavelength. As I tune the laser, the light does not move smoothly along the screen but winks out at one position and reappears at another—concrete evidence of the quantum jumps that give photons of certain wavelengths. The invisibility of the laser beam lends an eerie element to its formidable intensity. It carries only a few watts of power, the equivalent of a dim light bulb, but because its photons are identical, the beam can be brought to a fine point. Its concentrated power burns through thin metal in seconds, an impressive effect when nothing is seen except the growing damage. My arms have been burned as I unwittingly walked through the unseen beam. The strange sensation of a pinprick of heat encourages me to move quickly before the damage becomes deep. Larger CO_2 lasers easily cut through thick steel.

I use my CO_2 laser to energize a second laser in my laboratory, whose light replicates the cosmic background radiation left over from the Big Bang. Infrared light from the CO_2 laser enters a tube filled with vapor from alcohol or other organic substances, which contains complex molecules of carbon, hydrogen, and oxygen. These molecules are excited by the light and produce stimulated emission at wavelengths of hundreds to thousands of micrometers—the same range found in the cosmic background light.

The versatility of lasers comes partly from the variety of their media, which extends beyond a choice of gasses. Solid media include ruby, special types of glass, and semiconducting materials. Semiconductor lasers are the smallest types; they can be fabricated in densities of thousands to the square inch. Laser action can also occur in a liquid solvent that carries a brightly colored dye that can be changed to select a desired wavelength. Each of these types of lasers generates coherent light, some with enormous power. Unlike incandescent light sources, whose thermal processes are relatively slow, most lasers can generate extremely brief flashes of light, because quantum jumps are nearly instantaneous. With these unmatched abilities, lasers have become essential for the demanding applications of light that appear later in this narrative, such as telecommunications and medical diagnosis.

Although lasers meet most needs for light with specified properties, there are other specialized sources. Ultraviolet and X-ray light presents particular problems: It cannot be made by hot bodies, because the short wavelengths require temperatures seen only inside stars. And even after considerable effort for military use, it has so far proven impossible to make an X-ray laser. Instead, X rays for medical and industrial application are generated in evacuated glass tubes. Electrons emitted by a hot filament are accelerated by a voltage to a high speed. They smash into a metal target, where their abrupt halt sends massive shudders through the electromagnetic web to make X rays. Their impact also raises electrons in the atoms of the target material to levels of higher energy. These emit additional X radiation as they return to their original orbits.

The most powerful source of ultraviolet and X-ray light is also the largest of light-producing devices, the synchrotron, invented and originally used as a research instrument that accelerated fundamental particles to high speeds. Its heart is an enormous evacuated ring, measuring up to thousands of feet in circumference. To make photons, the ring is filled with electrons that circulate around it at nearly the speed of light. The rapid circular motion continually distorts the electromagnetic lines of force that connect each electron to the rest of the universe, creating a great outpouring of radiation. Each of the several dozen synchrotron sources scattered around the world is a center for research with light.

At the other extreme, there is light made in minute quantities or by less imposing means. No smaller sources can be imagined than those that coax a single photon from a single atom. Other sources emulate the quiet process of bioluminescence, which causes certain living organisms to emit light: Fireflies in a summer evening and life forms in tropical waters glow from reactions that change chemical energy into light without heat. The oxidation of a compound called luciferin causes the flash of the firefly, which enables male and female to find each other. For humans, the appeal is the usefulness of this efficient cold light. Chemical reactions animate the flexible plastic tubes filled with colored light made for children's play and the luminous sticks that create portable light without batteries. One ill-understood source of light, first seen in 1934, may also someday prove useful. It is sonoluminescence, in which high-frequency sound focused into certain liquids makes light. The conversion of sound into light seems to be connected to the formation and collapse of bubbles in the liquid, but it is puzzling that the light appears in flashes a million times shorter than would be expected from the properties of the sound.

Visitors to the 1893 Columbian Exposition, in their innocent enjoyment of electrical light, might have echoed the naive serenity of contemporary scientists about the nature of light, before the photon changed it forever. Still, knowing nothing of electrons, photons, and spontaneous emission, the cave painters, the Argands, Davys, and Edisons made light. Now deeper physical knowledge has given us more powerful and varied light. Its successful application requires that we also set its wavelength, strength, duration, and direction. Natural light, too, must be directed, focused, or filtered to give the colors of stained glass or to enlarge the image of a distant planet or a small object. Like the making of light, these manipulations began with ancient arts—the lore of dyes and pigments, early explorations of mirrors and shaped glass to guide light rays. We now control light with matter in subtle ways that direct photons or waves at the microscopic level of atoms and electrons, but light itself remains dual, intangible, and mysterious.

6

Shaping Light

... the "cold fire" of ruby—its deep red luminescence—
adds to the mystique and aura. . . . It is crystalline Al_2O_3
(sapphire) with a dilute doping of Cr_3^+ ions. . . .

—G. F. Imbusch and W. M. Yen, *Lasers, Spectroscopy,*
and New Ideas

Pure light was born long ago in the Big Bang, and it still travels
throughout the universe. Once light steps into the human world—
whether it comes from a distant star or from a struck match—it interacts with matter gaseous, liquid, or solid, each of which changes it.
Our vision confirms the interaction, for we see light as its photons are
absorbed in our retinas. The blue of the sky and the tints of the sea are
set as impinging light passes on its energy to molecules of air and
water; in fact all the colors of the world, every alteration of its light
rays, come as matter selectively reflects and absorbs, refracts and scatters light. There is an immense variety of visual properties within any
of the three great classes of matter—the clarity of glass and the lus-
trous opacity of gold, the neutrality of water and the sheen of mercury,
the limpidity of air and the poisonous yellow-green of chlorine gas.

This diversity testifies to the nearly infinite ways that matter interacts with light.

Early cave painters saw intense tints of red, brown, orange, yellow, and black in certain natural substances. From these, they made long-lasting paints to help them display their vision of the world, without knowing that the blood-red pigments contained iron oxide or that the blacks included carbon as charcoal, and without knowing why each material took on its color. The eleventh-century makers of Europe's oldest stained glass windows, in Augsburg Cathedral, colored them from ruby red through purple but did not know of the baffling complexities of glass or why impurities such as cobalt alter its tint.[1] The blue of sapphire, the red of ruby, the fire of the diamond—all had been coveted for ages, yet admirers did not know that sapphire and ruby are the same compound of aluminum and oxygen, differing only by the merest pinch of impurity that changes their color, or that clear scintillating diamond is made of the same carbon atoms as the black charcoal of the cave painters.

This empirical knowledge is now summarized in scientific principles that we confidently use to shape light. The lenses of cameras and telescopes, as well as the glass optical fiber that carries data throughout the world, use the same refraction of light that makes colors sparkle in faceted diamonds, but under precise control. And knowledge of the quantum gives new ways to shape light with matter, by converting photons to electrons and back again. That modern understanding grew in slow stages beyond trial-and-error learning. First came laws of optics that depended on the wave nature of light and its speed within a material. Before there was any inkling of the quantum physics of matter, those laws explained natural effects—the rainbow, the blue of the sky, the white of a cloud—and were used to manipulate light. Among their discoverers appear again the great explorers of light: Euclid, Descartes, and others; the mathematician Willebrord Snell; the astronomer Johannes Kepler; and the physicist Lord Rayleigh, perhaps the last gentleman scientist.

Their findings could later be placed into the unified framework of

Paramount Pictures, *Days of Heaven*, copyright 1978. Still image from motion picture.

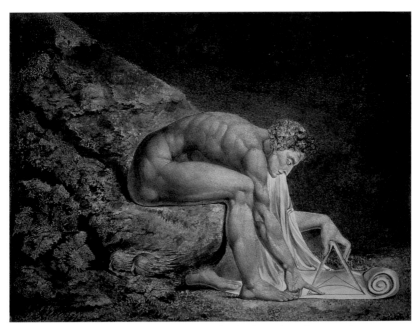

William Blake, *Newton*, 1795. Color print finished in ink and watercolor, 46 x 60 cm.

Katsushika Hokusai, *The Great Wave off Kanagawa*, 1823-1831. Woodblock print, 259 x 378.5 cm.

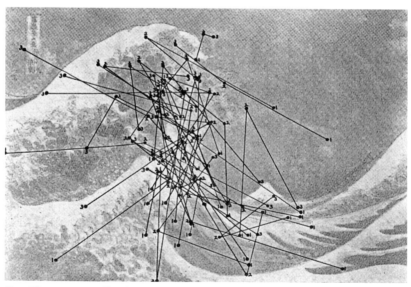

Guy Thomas Buswell, *The Great Wave* analyzed in *How People Look at Pictures: A Study in the Psychology of Perception in Art*. Plate XIV.

Pablo Picasso, *Portrait of Ambroise Vollard*, 1910. Oil on canvas, 92 x 65 cm.

Edward Hopper, *New York Movie*, 1939. Oil on canvas, 81.9 x 101.9 cm.

Edward Hopper, *Night Hawks*, 1942. Oil on canvas, 84.1 x 152.4 cm.

Georges de la Tour, *The Penitent Magdalen*, c. 1640. Oil on canvas, 134 x 92 cm.

Edgar Degas, *Le Café-concert des Ambassadeurs*, 1876-1877. Pastel on monotype, 37 x 26 cm.

Sonia Delaunay, *Electric Prisms*, 1914. Collage on paper, 49.84 x 33 cm.

Vincent van Gogh, *The Langlois Bridge at Arles*, 1888. Oil on canvas, 49.5 x 64 cm.

Vincent van Gogh, *Self-Portrait Dedicated to Paul Gauguin*, 1888. Oil on canvas, 59.55 x 48.26 cm.

M.C. Escher, *Hand Holding Reflecting Sphere*, 1935. Lithograph, 32 x 21.5 cm.

Gwyn Williams, *National Synchrotron Light Source*, March 1996. Photograph.

Vincent van Gogh, *The Starry Night*, 1889. Oil on canvas, 73.7 x 92.1 cm.

Hubble Space Telescope, Gaseous Pillars in M16—Eagle Nebula, April 1, 1995. Photograph.

Maxwell's electromagnetic theory, which described light in matter as well as in space. The theory, however, does not explain the origin of the properties of light in matter—why it travels faster in water than in glass, why it is reflected by gold but transmitted through air. The explanations are simplest for gasses, whose optical properties come from the quantum mechanics of single atoms or molecules. Liquids and solids contain myriad atoms and electrons that influence each other. Their linked behavior requires an extended quantum theory; it is the basis for condensed matter physics, the twentieth-century science that treats solids and liquids, including their optical characteristics.

In the late nineteenth century, however, scientists could describe much of the interplay of light with matter by classical theory, and artists could express that interplay through their pigments. These have been used since the days of the cave artists to paint surfaces with lasting tints. Small color-bearing particles are mixed with a liquid carrier, generally linseed oil for modern oil painting, which dries to leave a layer of color. Some ancient paints and dyes (pigment particles do not dissolve in paint, but they do in dye) carry wondrous histories. The red colorant carmine—in one form used in ancient Egypt and noted in the Old Testament, in another brought to Europe through the Spanish conquest of the Aztecs—is made from the crushed bodies of insects. It once dyed the red coats of the British army.[2]

Fascinating as these tales are, other pigments are deeply connected to the science of light. Many appear in the fiercely expressive color of Vincent van Gogh, which drives at us from the concentrated blaze of his work completed during the last five years of his life. In that time, at Arles and elsewhere, he painted all but a hundred of his nearly eight hundred canvases and formed the colors that touch us. "The more ugly, old, vicious, ill, and poor I get, the more I want to take revenge by producing a brilliant color . . ." he wrote in 1888.[3] The pigments Van Gogh used to make that brilliance, their brush-stroked thickness like the very congealed essence of color, bring together artistic expressiveness and scientific understanding of light and matter.

Van Gogh's art works show the array of pigments available in his time. His monkish *Self-Portrait Dedicated to Paul Gauguin* of 1888 has been extensively studied, because it has been damaged and repaired. It contains more than a dozen pigments.[4] Some—chrome orange and chrome yellow, emerald green, cobalt blue, ultramarine blue, zinc white—were found or invented in the eighteenth and nineteenth centuries; others—vermilion, lead white, charcoal black—are far older. Many of Van Gogh's paintings are illuminated by shades of yellow: the glowing yellow-orange of *Sunflowers,* the venomous yellow of *Night Cafe* ("one of the ugliest [pictures] I have done," he wrote[5]), the "fresh butter" color of *Yellow House.* He often used chrome yellow and the pigment cadmium yellow as well. Pigments containing cadmium were also popular with the impressionists. Monet favored cadmium yellow in his *London* series and many other works. Henri Matisse, whose striking colors were central to the fauvist movement, put cadmium yellow and cadmium orange into works such as *Bathers by a River* (1916–1919).[6]

Cadmium yellow is the chemical compound cadmium sulfide, whose remarkable history began with an accidental discovery. The metal cadmium was found by serendipity in 1817, as a by-product of zinc processing. It was not then known why its combination with sulfur should be yellow, but the "beauty and ... fixity ... of color" of that compound soon made it an artist's pigment, and a most flexible one. The quality of the yellow can be changed by varying the size of its particles or by adding zinc. When mercury or the metallic element selenium is added, the color moves toward the oranges and reds.[7] These shifts carry scientific as well as artistic meaning. The compounds that produce them are a family of semiconductors, materials that I and other scientists study because they can change light into electricity. This extended meaning for cadmium yellow, orange, and red was undreamt of in the time of Van Gogh or Monet. Now quantum theory explains how these semiconductors affect light.

Other pigments are also linked to the modern science of condensed matter. Considering only those types I have examined in my laboratory, there is the vivid vermilion and pure zinc white of Van Gogh's *Self-Portrait*. The first is mercuric sulfide, another kind of semiconductor. The second is zinc oxide, in scientific parlance an electrical insulator; it is intrinsically colorless but appears white in small particles. A different scientific issue also hovers around pigments. The cadmium, chrome, mercury, and lead in pigments are highly toxic metals. Some of the ills suffered by artists have been ascribed to toxic paints. One recent theory has it that Van Gogh's physical and mental afflictions came from an inherited disorder triggered by the paints he was said to ingest, although the theory blames the turpentine used for thinning rather than the metals.[8]

Artists ignored these scientific undertones as they covered canvas with paint, manipulating light with pigments so as to properly reflect it to the viewer's eye. Van Gogh painted a variety of natural effects heightened by his use of textured impasto to reflect light in arresting and expressive ways. His *Wheatfield with Cypresses* (1889) shows a subtly varied blue sky with thick swirling clouds, a tawny field, vegetation of light to dark green, foreground patches of red and white flowers. *The Langlois Bridge at Arles* of 1888 adds water and man-made materials. Its small bridge has a movable center section between stone abutments. A cart has just passed over, and a woman with an umbrella is crossing the bridge, which spans a canal that indistinctly reflects it and a pale blue sky with hints of clouds. Two tall cypress trees and other vegetation, as well as houses with white walls and pink or red roofs, complete the scene.[9]

Van Gogh saw the scene around the Langlois bridge by reflected light, which is how we usually see what surrounds us. Reflection occurs at the boundary between any two media. When light traveling in one medium encounters another, some photons enter the new material and some recoil from the interface, just as a tennis ball hurled at the surface of the court bounces into the air. Van Gogh's direct view of the bridge represents photons from the sun rebounding from wood and stone directly into his eyes. The image he saw in the water represents twice-reflected light: Photons rebounded first from the bridge to the surface of the canal, and again from that air-water boundary to the artist's eye.

The amount of light reflected depends on the properties of the two media, but the geometry of reflection is always the same. It was specified around 300 B.C.E. in the book *Catoptrics,* ascribed along with the book *Optics* to the mathematician Euclid. These founding works of geometric optics analyzed the straight-line behavior of light. The approach would have been natural to Euclid, whose classic *Elements* establishes the principles of plane geometry. Generations of students have learned Euclid's proofs about lines and angles, each flowing inescapably to the final acronym Q.E.D., *quod erat demonstrandum,* "which was to be proved." Rectilinear rays of light gave rise to Euclid's law of reflection. Van Gogh would have seen the law expressed if he had painted the sun's reflection in the canal: The line of sight from his eye to image, and that from image to sun, would lie at the same angle to the surface of the water. Followed back along its apparent line, the reflected ray would seem to come from a solar twin that lay as far beneath the surface of the canal as the true sun lay above it. The universe within a mirror extends exactly as far behind it as does the reality in front of it.

There is a complication, however, for the water in the painting is rippled, not smooth. Light rebounds unpredictably from such an uneven surface. Each separate ray follows the law of reflection, but as each comes from a different part of the surface set at its own angle, it emerges along its own direction. Van Gogh could see some of the light reflected from the forward slope of a ripple, but none from a back slope. The result is the distorted, incomplete reflection of the bridge that he painted. Such diffusely reflected light is said to be scattered, for it reappears along directions related indirectly or randomly to its initial line of travel.

Few surfaces found in nature are smooth enough to give a fine reflection. Only a perfectly flat surface does so, as Ovid recognized in his telling of the myth of Narcissus. That beautiful youth doted on his image in a pool "silver with shining water . . . whose glass no bird, no beast, no falling leaf had ever troubled. . . ."[10] Mirrors for daily use replace the flat silvery surface of a quiet pond with the flat high sheen of metals. Burnished metal was used in biblical times, but soon dimmed from wear and oxidation. By the Middle Ages, a liquid mixture of mercury and tin, or a thin metal foil, was applied to the back surface of a glass plate that protected the metal. The modern method of silvering that deposits a thin silver film on glass dates back to 1835, long before it was understood why metals reflect light so well.[11]

Even a smooth mirror forms a warped image if it is curved. The Roman statesman and philosopher Seneca the Younger wrote of one Hostius Quadra, who lined a room with curved mirrors to make provocative images of his lustful activities.[12] Curved mirrors entered into anamorphic art, whose most famous example is *The Ambassadors* (1533) of Hans Holbein. This richly detailed portrait of two visitors to the court of Henry VIII shows an unrecognizable object near their feet. A viewer walking past the painting sees that image become a human skull, whose perspective Holbein had sharply deformed. Other anamorphic art required a cylindrical or conical mirror to translate an obscure swirl of color and line into an intelligible reflected image. Mirror anamorphism was seen in Chinese art before 1619 and in European art throughout the seventeenth century. By the eighteenth century, the technique was appearing in popular etchings. Its coded images have ranged from a Caravaggio-like *Saint Jerome Praying,* to erotic scenes.[13] Curved mirrors themselves appear in art, from Jan Van Eyck's *The Arnolfini Marriage* of 1434, with its domed mirror on a wall that may reflect the artist himself, to M. C. Escher's meticulous *Hand Holding Reflecting Sphere* (1935) that indeed shows his image and puts two worlds into one picture.

The degree of reflection from a good mirror is independent of wavelength, but reflection can also select color if certain photons are absorbed. The red roofs in Van Gogh's *Langlois Bridge* absorb shorter wavelengths and reflect longer ones. The vegetation he painted contains chlorophyll, the compound that begins the food chain by using the energy of light to convert water and carbon dioxide into sugar.

Chlorophyll selectively absorbs photons at the blue and red ends of the spectrum. The remaining photons, mostly green, are scattered from the leaves of Van Gogh's cypresses, defining their color. The red flowers and tawny grain in *Wheatfield with Cypresses* contain other pigments that deal selectively with light, sometimes for biologically important reasons such as the attraction of pollinating bees. When Van Gogh painted these colors of nature, he was expressing pigment through pigment.

Although much of the natural world is seen by diffuse scattering from surfaces, scattering is at its most spectacular when a ray of light encounters motes of dust or molecules of gas. The stream of photons making up the ray is like a stream of water from a hose, which affects only what is in front of it. A bystander off to one side does not become wet unless an object placed in the flow deflects the water. Similarly, a ray of light in vacuum cannot be seen from the side, for photons register only on retinas in their line of travel, but a beam flashing past becomes visible if it is scattered by small particles. I use this effect to dramatically display the narrow pencil of light from a laser by clapping together over the beam two blackboard erasers full of chalk dust. That makes a swirl of motes which glow as they drift down through the light, scattering it to all sides.

Van Gogh's paintings show strongly scattered light, for blue sky and white clouds come from two versions of the process. The color of the sky was explained by the British physicist John William Strutt, Baron Rayleigh. His family had expected him to live as a country gentleman; instead, after he graduated from Cambridge University, toured the post–Civil War United States, and traveled up the Nile, he built a research laboratory at his ancestral home. There he remained, except for a time as professor at Cambridge beginning in 1879, when he filled the position left vacant by the death of James Clerk Maxwell. His major work is the lengthy *Theory of Sound,* but his hundreds of scientific papers touched all of classical physics, including an analysis of radiation from hot bodies that prepared the way for Planck's seminal contribution. He received the Nobel Prize for physics in 1904.[14]

Rayleigh showed in 1871 that when a light wave encounters a particle smaller than its wavelength, much of its energy scatters off in directions other than its original line of travel, with an efficiency that depends strongly on wavelength. Blue light is scattered ten times as effectively as red. As a beam of sunlight containing all the colors travels high above the Earth, it is scattered by air molecules. Blue light pours out laterally from its line of motion, much of it directed downward. We look up to see a sky filled with blue, whereas on the airless moon, the heavens are black. The same process explains why the world looks red at sunset. Late in the day, the rays of the low-lying sun traverse more of the atmosphere than they do at noon. They steadily lose blue photons as they go, gaining a distinct reddish cast in direct vision. Scattering is also responsible for the white of a cloud, which is a collection of water droplets, an airborne fog. Sunlight that encounters those particles is scattered to the side, but its color is not changed because the droplets are much larger than the wavelengths of light. Their scattering effect is like reflection from a horde of small curved mirrors.

Natural scattering is mimicked in the arts of oil painting and stained glass. In the pigment zinc white, particles of zinc oxide larger than the wavelength of light scatter white light as do droplets in a cloud. The same effect occurs with ordinary clear glass, which appears white in ground form. In other cases, light is altered to artistic effect when it is scattered from particles smaller than its wavelength. This is what changes the shade of cadmium yellow according to the size of the particles of cadmium sulfide. And when ultrasmall particles of gold are imbedded in stained glass, they strongly scatter the blue out of sunlight and impart a lovely red to the transmitted light.

That transmitted light reminds us that although we usually observe only the surfaces of the world by reflected light, we sometimes see deeply when light travels through a layer of matter. Some materials are opaque, readily absorbing photons. Light does not reach the far side unless the layer is extremely thin. In other media, photons travel with little or no loss, and many emerge even from a thick layer. If they come out undistorted as well, still carrying their original image, the material is transparent, like glass. If they emerge diffused, with no clear image, the material is translucent, like alabaster or fine porcelain.

Light transmitted is less commonly painted than light reflected, but some forms of art rely on it. When watercolorists use the transparent form of their paint, light penetrates the layer of color and is reflected back to the viewer from the underlying white surface. This gives a luminous quality that John Singer Sargent called "captured light." Stained glass comes alive only as light streams through a cathedral window or glows through a Tiffany lampshade. And cinematography becomes meaningful to the viewer only by means of light transmitted through moving film.

To understand transmission, it must be viewed from within the medium in which it occurs. Suppose Van Gogh were to paint the underwater world of his canal near Arles from a diving bell, perhaps one of Captain Nemo's ornate Victorian devices. He would see sunlight transmitted through the water and would observe that the light became dimmer as he descended. Photons are absorbed by water, a greater number disappearing as a greater depth of water intervenes. Their energy becomes heat, which is why water becomes warm under the sun. And as Van Gogh went deeper, the light would become greener. That subaqueous look comes because water absorbs red light more readily than it does green, blue, and violet. Like light reflected, light transmitted can be selectively colored. Clear glass and plastic are even-handed in their treatment of photons, but they can be altered to absorb all photons except red ones, for instance, which pass through. Such selectivity makes colored filters for theatrical or photographic applications, tinted lenses for sunglasses, and the red-and-blue spectacles once used to view three-dimensional films.

Looking up from beneath the water, Van Gogh might also see the sun displaced from its true position. Had his painting shown that effect, it would have illustrated refraction, another influence of matter on light. It occurs as a light wave moves into a new medium in which it has a different speed. Earlier I likened the crest of such a wave to a line of dune buggies sweeping at an angle from hard sand to soft, or a shoulder-to-shoulder row of soldiers tramping obliquely from asphalt to plowed field. The abrupt change in speed bends the light ray at the interface. Refraction is easily seen in water, where light slows to only 124,000 miles per second. The effect causes the apparent bending of a partly immersed object, and it is responsible for the rainbow, which appears when raindrops separate light into colors that can be seen from certain vantage points. And because refraction also occurs in glass, it is the reason that lenses focus light.

The effects of refraction on light can be concisely described. If light enters water vertically, the entire wave crest slows at the same instant and there is no bending. At any other angle of entry, the ray is bent in the water so as to travel more nearly at right angles to the interface. This is true as light passes from any medium into one where its speed is lower. Conversely, if light streams from water to air, it is bent to lie more nearly along the water-air boundary. At certain angles, there is a surprising outcome: Light originating within the water just skims along the interface or returns to the water, never penetrating into air. This confinement, called total internal reflection, can occur whenever light attempts to move from any medium of low speed to one of high. It is responsible for the compelling luster of a diamond cut in the so-called "brilliant" style, where the facets are formed to capture light within the gem.

The angle of refraction depends on the angle of entry and the speed of light in the media, but the exact mathematical relation took more than a millennium to uncover. The search began with Claudius Ptolemy of Alexandria, who examined refraction around 130 A.D. He is best-known for his astronomical observations that elaborated Aristotle's earth-centered vision of the solar system. Like many other astronomers, Ptolemy also studied optics. He tabulated results for different angles of approach and of refraction as light encountered water. But in an aberrant application of Greek ideals of beauty, he is said to have altered the data to conform to an aesthetic standard he "knew" to be right.[15] Aesthetic considerations have sometimes led to mathematical equations that express meaningful scientific ideas, but they are a poor reason to manipulate data. And even when data are accurately reported, that is not enough; they must also be analyzed to give a workable theory or equation.

That reduction began in the early seventeenth century with the German astronomer Johannes Kepler. He is best known for his analysis of the data gathered by the great Danish astronomer Tycho Brahe, which showed that the orbit of Mars is an ellipse. Ptolemy had assumed that planets moved in circles, and Kepler found his own result difficult to accept, but elliptical orbits became one of his laws of planetary motion, which supported the heliocentric system set forth by Copernicus. Newton later derived those laws from the action of gravity as it held the planets around the sun.

Although Kepler's work opposed the Ptolemaic view, he echoed Ptolemy in considering the process of refraction. In 1611 Kepler published an approximate relation between the angles of the incoming and the refracted light, which depended on the speed of light in each of the media. The role of speed is summarized in the quantity called the refractive index of a medium, defined as the speed of light in a vacuum divided by the speed of light in said medium. Since light always moves more slowly in a material than in a vacuum, the index is always greater than one: It is 1.3 for water, corresponding to a speed of 124,000 miles per second; the value varies from 1.5 to 1.8 for different glasses, and for diamond it has the extremely high value of 2.4.

The complete, accurate expression for the refracted angle came from Willebrord Snell, professor of mathematics at Leiden, who maintained a variety of scientific interests. He invented the method of triangulation, in which three reference points allow an unambiguous calculation of location on the surface of the earth, and calculated the non-repeating digits in π. Physicists know him for Snell's law, the exact statement of refraction he formulated in 1621 though it was not published at that time. Descartes did so sixteen years later, after he apparently derived the law from his theory of light as pressure transmitted through a fluidlike substance. (The issue of priority of discovery versus priority of publication still lingers in the intense world of scientific research.)[16]

Without Snell's crucial equation refraction would remain only a bit of optical magic; with it, refraction becomes a tool to form light in a lens. Imagine Newton's prism in its triangular cross-section, with one side—the base—set horizontally. As a ray of light enters the glass, it is refracted downward toward the base and down again as it leaves the prism. Now press the base of another prism against the base of the first, making a diamond-shaped cross-section. A ray entering the bottom prism is bent upward at the first and second interfaces. Inevitably it meets the top ray. The same two prisms, placed point-to-point rather than base-to-base, form an X-shaped cross-section. Then rays in the top and bottom prisms are bent up and down respectively, to diverge rather than converge. These are prototypes of true lenses, circular pieces of glass shaped to refract light as the combined prisms would. A convex lens, thick at its center like the diamond-shaped arrangement of prisms, brings the rays in a beam of light to a common focus; a concave lens, thin at its center like the X-shaped arrangement, diverges a beam of light so it spreads out on its far side.

Snell's law of refraction makes it possible to calculate with assurance the focal point and other properties of a lens, for optical applications. Every child who has played with a convex lens knows that sunlight concentrated at its focal point can set things alight. This has been understood since the time of Aristophanes, who spoke of a burning glass in *The Clouds* of 424 B.C.E.[17] And a convex lens magnifies the image of a small object placed near its focal point, from which it sends an enlarged sheaf of rays to the eye. Seneca knew that objects appeared larger when seen through a water-filled globe, which approximates a convex lens, and the Romans apparently used true magnifying glasses.[18] The magnification by a single lens is relatively modest but brings the microscopic world significantly closer. The seventeenth-century Dutch naturalist Antonie van Leeuwenhoek revolutionized biology with lenses that he ground to a high standard of perfection; some could magnify nearly three hundred times. With them he found minute bacteria, single-celled protozoa, and red blood cells.

A single lens can also correct faulty vision. The emperor Nero is said to have used a curved piece of emerald on a ring to help his poor eyesight, but spectacles that held lenses before the eyes did not appear until the thirteenth or fourteenth century. Their invention has been ascribed to the scientifically minded Franciscan monk Roger Bacon, but it is more likely that his writings inspired other unknown inventors. Bacon championed the modern view that observation and measurement of the natural world are the proper foundation for science. He wrote on varied scientific subjects, but he also accepted astrology. His knowledge of optics had been influenced by the work of the Egyptian Ibn al-Haytham or Alhazen, and he understood to some extent how light traveled through lenses. Bacon speculated that these might correct defective sight, and "cause the sun, moon, and stars to descend here below. . . ."[19]

Held before the eye or floating on the eyeball (the British astronomer John Herschel proposed the contact lens in 1827), an artificial lens helps to focus light on the retina. The living convex lens at the front of the eye can be made more or less curved by surrounding muscles, which is how we focus on objects near as the morning newspaper and distant as a star. But in some imperfect eyes, the distance from lens to retina is too large or too small to be properly bridged by the projected image. Then remote objects, or near ones, cannot be clearly seen. Even in a correctly formed eye, the lens loses flexibility with age and can no longer focus on a near point. These difficulties are corrected by a concave lens that diverges incoming rays to match a long nearsighted eye, or a convex lens that more rapidly brings rays to a focus that matches a short farsighted eye. Corrective lenses are the most widespread optical devices; half of all Americans use them.

A single lens, however, is inadequate for other applications. It was the breakthrough idea of one lens acting on the image made by another that underlies the powerful compound microscope and telescope.

Both seem to have been invented nearly simultaneously in Holland about 1600. The Dutch spectacle maker Hans Lippershey applied for a patent on the telescope in 1608, supposedly after accidentally observing the effect of a convex and a concave lens placed in line. Word soon reached Galileo Galilei in Padua, and by 1610 he was scanning the heavens through the new device. As he made discovery after discovery—observing four moons of Jupiter, resolving the luminous band of the Milky Way into stars—he began the modern study of cosmic light. The compound microscope may have been invented by the lens maker Zacharias Jannsen. Robert Hooke's *Micrographia* of 1665, which provided evidence for the wave theory of light, contains many observations made through such a microscope, including the discovery in plants of the structures he called cells.

The telescope and the microscope each use two lenses set in a tube, with a convex lens placed nearer the object to be viewed. In the microscope, that lens is like a magnifying glass applied to a small object near its focus. The resulting enlarged image is enlarged again by the second lens, which presents to the eye a final image magnified a thousand or more times. In a telescope, the two lenses create a final image which looms larger on the retina than that from the unaided eye, bringing the object apparently nearer.

Early versions of these devices had a serious problem inherent in the refraction that defined them. Newton had used his prism precisely because it separated light into colors. Each wavelength has a different speed in glass and is bent through a different angle, according to Snell's law. This creates the scintillating colors of a diamond: As light flashes from facet to facet inside the gem, the high index of refraction separates it into clearly discernable colors. But the refractive division of wavelength is undesirable in a lens, for it brings each color to a focus at a slightly different point. Instead of a sharply defined spot of white light, there is an elongated blur of separate colors. The lens cannot form a clear image. If you wear eyeglasses with thick lenses, you may see this chromatic aberration as a shading of color along the edges of objects.

Newton had tried to eliminate the effect when he built a telescope with lenses, but he was unable to do so. He designed instead the reflecting telescope, which replaces lenses with mirrors. A concave mirror, thinner in the center than at the edges, directs light to a central point like a convex lens; it can focus sunlight into a spot of blazing radiation. An old and unlikely tale has it that Archimedes defended Syracuse during the Second Punic War by burning Roman ships with sunlight directed by mirrors—unlikely given the mirrors of the time and the fact that the ships would have to obligingly hold still long enough to ignite.[20] But curved mirrors are now used in furnaces where focused sunlight melts metals. And as Newton realized, there is no chromatic aberration in a mirror-based telescope, for each color is reflected and focused equally. Such telescopes remain important astronomical tools.

Devices that require lenses, however, must eliminate distorted color if they are to give true images. The London optician John Dollond found the answer in 1758. He used two kinds of glass with different refractive indices and therefore with different patterns of chromatic aberration: the older type called crown glass, and the newer flint glass, invented in 1675. Now called lead glass, it contains lead oxide, which increases its refractive index. This gives it a diamondlike luster when it is cut with facets—the reason that Waterford crystal has long been prized. Dollond combined a concave piece of flint glass with a convex piece of crown glass into a single lens that superimposed their patterns of color into well-focused nearly-white light. Later, Dollond's son Peter extended this doublet arrangement with a third crown-glass lens, which corrected the residual aberration to produce yet a better image.

Manipulation by multiple materials reaches new heights in modern lenses. Compared to Dollond's two optical glasses, there are now dozens of types, each with its own exactly known refractive index. With computers to carry out the complex refractive calculations, these varied optical qualities are routinely combined into superb lenses. A good camera lens contains seven or more elements made of different glasses to give excellent color-free focusing. And the choice of glasses is supplemented by plastics, as found in soft contact lenses and single-use cameras. Enhanced control of matter also allows more effective use of light. For instance, clearer images result when stray reflections are removed from the front surface of a lens, which can be done with a transparent coating that produces destructive interference. The fabrication of the requisite extremely flat and thin films is a relatively recent achievement. Most other common means to shape light—dyes for photographic and cinematographic color film, spotlights and filters for theatrical illumination—use twentieth-century methods and materials to form nineteenth-century light.

Classical theory also unified the piecemeal understanding of light in matter that had grown over millennia. Maxwell's electromagnetic equations explained Euclid's law of reflection and Snell's law of refraction, which are inherent in the behavior of waves at an interface. The equations predict how much light is reflected from a surface or transmitted through a material—and they explain the blue of the skies. When the oscillating electric field of an electromagnetic wave encounters electrically charged particles, such as the atoms in air molecules, it sets them vibrating, to create new light at the same frequency. Maxwell's equations show that some of this light emerges at right angles to the direction of the original light ray and is more intense at shorter wavelengths, neatly summarizing all the features of Lord Rayleigh's scattering.

The image of charged particles set vibrating by light also begins to explain the ways of light in liquids and solids: why it travels more slowly in matter than in vacuum, why it is highly reflected by a metal. Light undulates in empty space independently of any masses, but as it engages the charged atoms that make up matter, it becomes linked to masses, which resist motion. Light entangled with matter is a combined electromagnetic and mechanical wave. It moves more slowly than light in space, like a water wave that loses speed as it drags along sand or pebbles. Infrared light affects the massive atomic nuclei, whereas the higher frequencies of visible and ultraviolet light engage the less massive electrons. Hence the speed of light in matter depends on its frequency. This gives a different refractive index for red light and for blue, the final step in explaining why Newton's prism split white light into colors.

The interaction of light with electrons also explains why a metal reflects nearly all incoming light. Any metal conducts electricity because it has a large number of electrons that can move freely. When light impinges, these electrons oscillate under the influence of its electric field and radiate new electromagnetic waves. The new waves are out of step with the incoming ones, creating destructive interference that cancels the light before it can penetrate deeply into the metal. Instead, most of its energy is reflected from the surface.

Classical theory, however, leaves unexplained many broad strokes and fine details that light paints in matter and that matter etches on light. Granted that gold and silver reflect light because each contains a swarm of free electrons, but why do these metals contain such electrons, whereas diamond does not? Why does gold carry the warm cast of the sun, silver the cool shading of the moon? Classical theory does not clarify why some small changes in the composition of matter largely affect light. Pure crystalline sapphire, two parts aluminum joined with three parts oxygen in the compound aluminum oxide (Al_2O_3), is beautifully clear. Add a few atoms of one kind, and the result is a crimson ruby; inject a few atoms of another kind, and a blue sapphire results. Finally, classical theory does not explain the colors of painter's pigments. Why is cadmium sulfide yellow, and why does it become cadmium orange and cadmium red as selenium is added?

It is tempting to seek answers in a simple theory of matter. Once that might have been the idea of the Four Elements—earth, air, fire, water—proposed in the fifth century B.C.E. by the Greek philosopher Empedocles. Setting aside fire, the idea has its modern counterpart in the categories solid, liquid, and gas that alter light. These are too broad to make acute distinctions, as between opaque gold and transparent glass, yet they give an important clue: The optical properties of matter, indeed all its properties, depend on its internal order, which differs in each category. The atoms of a gas are in rapid random motion, with no semblance of internal structure. A gas has no shape except that imposed by its container. In contrast, the atoms of a solid are linked by forces that hold them fixed relative to each other. A stone or a gold ingot maintains a definite form without external support. Somewhere in between lies a liquid. Its atoms are connected tightly enough to keep them from flying into space (although some may evaporate) but loosely enough that a liquid takes the shape of any container.

These structural characteristics define the quantum physics of each kind of matter. The lack of structure in a gas makes its analysis the most straightforward. From the Bohr atom of early quantum theory to the Schrödinger equation for the wave function, our quantum understanding is highly developed for isolated atoms and molecules. This approximates conditions in a gas, where the atoms or molecules do not interact except for occasional collisions. The same energy levels involved in discharge lighting or the operation of a laser specify which photons a gas absorbs. This explains the clarity of oxygen, the ultraviolet behavior of hydrogen, and the infrared absorption of carbon dioxide.

Explanations are more difficult for matter condensed into solids and liquids, the final coming together of elementary particles these billions of years after the Big Bang made the first scraps of matter. The close-bound atoms in these condensed systems influence each other, creating properties different from those of isolated atoms. So astronomically large is the number of atoms and electrons, so intricate their interactions, that full comprehension of solids and liquids would seem impossible. Some types of solids, however, possess a particular physical structure that simplifies quantum understanding. These are the crystals, the atoms of which lie fractions of a nanometer apart in a fixed and ordered spatial pattern. The pattern is different for each type of crystal but always consists of a basic unit repeated over and over, as the solid bulk of a hotel could be constructed by stacking identical modular rooms along its length, width, and height. That three-dimensional lattice of atoms recurring in space is at the heart of crystalline behavior.

Gold, diamond, and cadmium sulfide are all crystals; glass and clay are not. Internal crystalline order is often apparent on the human scale. Sodium chloride, or common salt, is the best-known example. If you could stand on one of its atoms, you would see sodium and chlorine atoms alternating in a three-dimensional checkerboard extending in all directions. That cubic symmetry is echoed in the shape of grains of salt. Other crystals are more complex, for instance with a honeycomb-like hexagonal format, as found in cadmium sulfide. The great variety and elegant symmetry of crystalline shapes has fascinated through the ages. The microscopic reality underlying that beauty was itself first explored by invisible light, when in 1912 X rays vividly displayed the atomic structure of crystals.

The properties of a crystal reflect its internal energy levels, which come from the quantum nature of its electrons—an extension of what occurs in a single atom. In an atom, the De Broglie matter wave for each electron fits exactly around its orbit to give constructive interference. This sets the energy for each orbit, giving a set of allowed levels separated by gaps in which electrons cannot exist. In a crystal, the matter waves of the electrons roll through the atomic geometry of the lattice, like watery crests and troughs among a regular pattern of pilings. Waves survive in that electronic sea only if their wavelengths match the dimensions of the lattice, so that they interfere constructively as they rebound from the atoms. Waves that do not match the lattice undergo destructive interference and disappear. Their missing wavelengths correspond to a range of missing energies, a so-called "band gap" that is like the space between two energy rungs in an atom. The energy levels above and below the gap are like a higher and a lower energy rung in an atom, except that they house enormous numbers of electrons. Electrons in the lower levels are embedded in the crystalline structure; those at the higher energies above the gap move freely through the solid.

An atom absorbs only those photons with enough energy to lift an electron to a higher level. Similarly, a crystal absorbs only those photons with sufficient energy to carry an electron from below the gap to above it. This immediately explains the color of zinc white: The zinc oxide in the pigment is a crystal with a band gap so large that even the most energetic violet photons cannot raise an electron across it. All colors travel unimpeded, and thus the material is transparent. Ground into pigment, particles of zinc oxide scatter white light without altering it. And since few electrons ever cross the gap by any means, zinc oxide is a poor conductor of electricity. In general, any of the limpid crystalline materials, such as pure quartz, is also an excellent electrical insulator.

The sparkling clarity of a crystalline insulator changes when an impurity is added, as quantum theory also explains. Less than one percent of the metallic element chromium added to transparent aluminum oxide makes it red ruby, whereas a trace of titanium turns it into blue gem sapphire. Each foreign atom that enters the host lattice behaves like a single atom with its own particular quantized energy levels that determine how the impurity deals with light. Chromium in aluminum oxide absorbs photons so the crystal looks red, whereas with titanium the color blue dominates. The same impurities determine the light emitted by the crystal if its atoms are excited. The quantum processes that make ruby beautiful also underlie the ruby laser that was invented in 1960.

A metal such as gold or silver differs from an insulator because there is no band gap to impede the motion of electrons. There are always many of them available to carry electrical current and to vibrate so that incoming light is reflected. The excellent electrical qualities and the high gleam of a metal are different aspects of the same quantum reality. The varied shadings of gold and silver, copper and platinum, come from finer details of the electronic energies, which differ from metal to metal. In gold, these determine that wavelengths from blue to green are absorbed to some extent, so that reflected light has a slight preponderance of shades from warm red to yellow.

Some crystalline materials have band gaps smaller than those of an insulator yet have fewer free electrons than a metal. These are the semiconductors, whose name alludes to their intermediate electrical properties. The best-known example is silicon, used in computer chips and every other kind of integrated circuit. Under certain conditions, electrons leap across the gap in a semiconductor and the material conducts electricity; under others, a semiconductor becomes an insulator or conducts electricity by the flow of positive rather than negative charge. These variations can be delicately controlled, making semiconductors the versatile manipulators of electrons that define much of our technology.

The optical consequences of the gap also make semiconductors flexible manipulators of photons and finally explain the colors of the pigments that Van Gogh and the impressionists put to artistic use. In cadmium yellow—that is, cadmium sulfide—the gap corresponds to the energy of a green photon. Photons of that or smaller wavelength have enough energy to raise an electron across the gap, so the colors from green to violet are absorbed. The remaining yellow to red photons determine the color of the material. If selenium is added, that makes a new compound with a smaller gap, which absorbs photons of lower energy and longer wavelength. As the dosage of selenium increases, yellow and orange diminish, and the color approaches deepest red. The colors of cadmium yellow, orange, and red, as well as those of vermilion and other pigments, directly express their quantum properties.

The absorption across the gap that causes such pleasing colors has technological value, for it transforms photons into free electrons. This conversion of light into electricity has become widely used because of the versatility of semiconductors—the varying band gaps of different types respond to different wavelengths of light. Cadmium sulfide efficiently absorbs sunlight and is made into solar cells that draw power from sunlight by changing it into electricity. Some semiconductors operate at the blue end of the spectrum; others, like the widely used compound gallium arsenide, have gaps corresponding to infrared photons. These are used to detect this invisible light, as its photons are turned into a measurable electrical flow.

The inverse transformation, electrons into photons, can also occur in a semiconductor. It is analogous to discharge lighting, in which electricity excites a gas and causes it to emit light. A piece of semiconductor is formed into a small device that is connected to a battery. That brings a steady supply of energetic electrons, which continually descend through the band gap to lower energies, like an electron dropping between levels in an atom. Each descent makes a photon with energy equal to that spanned by the gap. Tiny infrared lasers, made on this principle, are at the heart of music reproduction from compact discs and of data transmission by light over optical fibers.

Noncrystalline or amorphous solids, with atoms arranged in no particular order, shape light as well. One example is carbon, which makes a black pigment when it is used in the amorphous form of charcoal. (This example points to the importance of structure; if the same carbon atoms are marshaled into a lattice, they give the crystalline transparency of diamond.) The most important amorphous material is glass, which is misnamed when it is called "crystal" glass. In principle, quantum theory describes glass and all amorphous substances, but this has proven difficult without the hard-edged geometry of a lattice as a guide to quantum behavior. Glass has shaped light for four thousand years, yet we do not understand it.

In the nineteenth century, Faraday alluded to the special properties of glass when he called it a "solution of different substances." This becomes apparent in the oldest form of glassmaking; ordinary sand (the compound silicon dioxide) is mixed with other minerals and melted into a hot liquid that solidifies as it cools. A hot liquid that cools slowly tends to crystallize, like a lake turning into ice as it gradually chills below freezing. Rapid cooling, however, creates the ambiguous material glass: rigid yet lacking microscopic order. With its liquid origins, glass has sometimes been treated as a highly viscous substance that slowly creeps under its own weight, as is said to occur in ancient cathedral windows that have become thicker at the bottom as the glass settles.

The transparency of glass derives from the behavior of electrons in individual molecules of silicon dioxide. But because the molecules are disordered, it is not yet possible to predict the optical properties of glass from a quantum model for all the linked atoms of the material. Even so, principles of chemistry and physics guide the development of glass in its great variety of refractive indices and colors, thermal and mechanical characteristics. Scientific principles have led to new types, such as the photochromic glass that changes its tint in response to light, but much of the forming of light by glass still draws on knowledge that grew over millennia of use, such as the tinting of glass by the addition of metal particles when it is molten.

The difficulties that defeat a full understanding of glass are even greater in liquids, in which the molecules are in motion. Molecules of water form a definite crystal lattice when the liquid is frozen into ice. When heat is applied they begin to move, and the ice melts as the lattice collapses. Even in the liquid state, the molecules do not move with complete randomness as in a gas but form and reform larger structural elements—a dynamic behavior so rapid and complex that a theory of water, or of liquids in general, has yet to be developed. Still, the optical properties of some liquids are understood. In many cases they derive from the behavior of solids, such as the particles of carbon that make ink when they are suspended in water. Some characteristics come from scattering: Milk is white because its suspended globules of butterfat scatter light as water droplets do in a cloud. Other liquids display intrinsic optical properties: A drop of mercury gleams because it contains many free electrons, as does a solid metal. But because liquids occupy a middle ground between the independent randomness of a gas and the constrained geometry of a crystal, the ways of light in that state of matter remain difficult to grasp.

The dualities of light, it seems, persist into its shaping by matter, which combines the certainties of crystalline solids with the ambiguities of liquids and of amorphous materials. The ancient material glass is itself poised between solid and liquid, transparent to vision yet opaque to understanding. But even if we knew amorphous substances as well as we do crystals, the wave-particle mystery of light remains, and matter brings its own quantum enigmas: the particle-wave duality of the electron, the transformations of photon to electron and electron to photon.

This does not exhaust all the rich complexity of light in matter, as Van Gogh's art shows. His paintings of earthly scenes portray light reflected from natural pigments and from water, and light scattered by air, but his pictures of night skies—*Starry Night over the Rhone* (1888), *Road with Cypress and Star* (1890), and most famous, his *The Starry Night* with its enormous incandescent sky looming over a church spire, painted at Saint-Rémy in 1889—add more. They extend his art from the world around him to cosmic scenes, and they show an artistic wonder—the capture of light from shining suns and planets in bits of colored pigment that can only reflect light, not create it. This consummate forming of light also carries scientific validity, for these pictures correctly represent the heavens. Except for some artistic license, the unusually close configuration of crescent moon, Venus, and Mercury in *Road with Cypress* and the Big Dipper in *Starry Night over the Rhone* are accurately depicted according to the time and place from which each scene was painted, and various features in *The Starry Night,* including its enormous central swirl of light, resemble known astronomical objects.[21]

The thick paint and strong brush strokes that defined Van Gogh's art toward the end of his life symbolically unite what we know about light: how it is altered on the microscopic scale of atoms and quanta, which defines how pigments color light; how it looks in the common world of wheatfield, bridge, and canal; and how light fills the cosmos, as in *The Starry Night.* In Van Gogh's paintings, as in all art, the controlled use of matter shapes light. The artistic bending of light that shows a leaf or a star has its counterpart in common and scientific uses of light, from the small scale that reads musical markings on a compact disc to the enormous scale that employs light at terrifying intensities and examines cosmic illumination. These draw on every means of making and forming light.

7

Using Light

We've come from the crazy to the impossible to the impractical.

—Alan Huang, pioneer in light-based computers

I sit at my personal computer and load a thin, shiny five-inch disc identical in appearance to a compact music disc. The platter spins, the software sets to work, and in a moment I can display on the monitor screen a French impressionist painting, chosen from among hundreds on the disc. I select another disc, and I can call up on the screen any work of art in the National Gallery of London. Light is the physical link between the data stored on the disc and the forms and colors so easily selected and displayed. Van Gogh portrayed cypress leaves and burning suns through bits of pigment; today those scenes are expressed through binary digits stored on a CD-ROM (compact disc–read only memory) and read by a laser, to control flecks of color on a monitor screen.

The uses of light range from the aesthetic to the mundane, and they extend to our study of the universe. These employ the full range of photon energies, from dangerous X rays to the weak light of the infrared. Many applications, but not all, involve computers and binary digits. Many, but not all, use novel sources, from small solid-state lasers to immense synchrotrons. All, however, weave separate elements into a system of light, just as the living visual system requires lens, retina, nerve channels, and brain. Human vision does not provide its own light, and this is also true of some artificial systems, such as the sensors that examine the Earth from orbiting satellites. But many systems require a source of a particular wavelength or extraordinary power or specific temporal character; a detector that changes light into electrons to determine its presence or measure its intensity; and a channel to carry light from source to detector, which may also modify the light as it passes from one to the other.

This general scheme applies to systems of light large or small, scientific or commercial. I have measured the impact of light on matter with a tabletop laser emitting a few watts of light and at a laboratory of light the size of a football field, one of the awe-inspiring synchrotrons that makes kilowatts of ultraviolet, X-ray, and infrared light pour out of orbiting electrons. In both small laboratory and giant synchrotron, the light is modified as it encounters the material under study. Appropriate detectors determine those changes, which give clues to the inner workings of matter.

The same systemic elements appear in the compact discs that store music or the CD-ROM that holds Van Gogh's *Wheatfield with Cypresses,* but the elements are executed in different scale and style. A minute semiconductor laser makes a few milliwatts of infrared light that is reflected from the shiny surface of the spinning disc. The light is sensed by a detector, also made of a semiconductor, until the reflection is interrupted when a dark area on the disc spins by. The detector turns those "on" and "off" blinks of light into pulses of current that echo the binary coding embedded in the CD-ROM. Carried via copper wires, the impulses enter the computer. There they control a beam of electrons, which charges each small region of the cathode-ray tube in the monitor with appropriate color and intensity. The result of this enormous paint-by-numbers task is the image of *Wheatfield* shining on my screen.

The CD-ROM illustrates one advantage of light as tool: Its tiny wavelengths allow each iota of data to be stored in a small area formed and picked out by a ray of light, allowing for extraordinary densities of information. That is the reason for the rainbow glitter of a compact disc. Adjoining segments of its miles-long spiral track, which contains the billions of light and dark areas, are so closely spaced that they reflect light according to wavelength. The next step, now under development, will be to store information at even higher densities in a three-dimensional pattern of light spread throughout a crystal.

The semiconductor lasers that make immense storage possible also animate the national and global systems of light that carry telecommunications, the Information Superhighway, and link computers through networks such as the Internet. In these webs, photons race through the microscopic glass alleyways of optical fibers, replacing the older system in which electrons stream through copper wires. It is not that electronic technology is suddenly obsolete—electronic computers still set records in speed and capacity as they manipulate huge clots of data—but for some applications, only light will serve, as in presenting information from a computer monitor. And light is replacing electricity wherever the insatiable desire for faster computers and more powerful communications is finally outstripping what electrons can do.

A century ago, when the electron was an exotic elementary particle, few could have foreseen an electronic technology and where it would lead. Although the photon is still exotic, new ways to shape light with matter are creating a technology of photonics, which may well define the twenty-first century, as electronics is defining the twentieth. Photonic technology is the latest stage in the transition from the Industrial Revolution to the Age of Information, which deals with reality in ever more refined ways.

In the early Industrial Revolution, the brute force of expanding steam turned wheels to perform mechanical tasks. Later, steam turned electrical generators, making a cleaner flow of power that drove motors and heated filaments. But moving electrons did more—they made a new technology of communications, from telegraph to television, and they made the computer possible. Charles Babbage, the nineteenth-century pioneer of computing, designed (but never completed) the extraordinary Analytical Engine, an array of gears and rods driven by a steam engine that displayed most functions of the modern computer. A mechanical computer, however, was too limited and too limiting. The computer became marvelous only when it used electrons rather than gears, which gave it speed, capacity, and versatility. Yet electrons are still limited, compared to photons. Information cannot be transmitted any faster than by light, whether in free space or in matter. Light travels through a transparent material with little loss of energy, whereas electrons make wasteful heat as they flow through wires. Most important, light can carry enormous amounts of information, because it vibrates at such high frequencies.

Frequency, the number of waves transmitted per second, is nearly the same as the capacity to carry information. Every communications system—from two tin cans at the ends of a string to a television broadcasting facility—has a bandwidth, the difference between the minimum and maximum frequencies it can transmit. The larger the bandwidth, the more information can be conveyed. Think of the rainbow's arc: Its colors from red to violet correspond to frequencies from 400,000 gigahertz (a gigahertz is a billion hertz) to 750,000 gigahertz; that is, they span a bandwidth of 350,000 gigahertz. If vision extended only from 400,000 to 450,000 gigahertz, the reduced bandwidth of 50,000 gigahertz would present a world in red—exactly what happens when you slip a piece of red plastic before your eyes, losing polychromatic richness and significant information.

The same idea applies to radio and television. The electromagnetic waves of AM (amplitude modulation) radio vibrate at hundreds of kilohertz. Each AM station is allocated a bandwidth of 10 kilohertz, a limited range that cannot transmit the full spectrum of human hearing, which extends to 20 kilohertz. The waves of FM (frequency modulation) radio, however, oscillate at megahertz frequencies. Every FM station is allocated 200 kilohertz, ample to accommodate all that humans can hear and carrying a richer sound than AM radio. The more complex information sent over a television channel requires twenty-five times the bandwidth of an FM station, and the proposed new high-definition television requires another factor of six to deal with its greater detail, for a total bandwidth of 30 megahertz. But these values are insignificant compared to the ultrahigh frequencies of light that offer immense scope for the transmission of data.

There is another way to look at frequency as linked to the most profound insight of the Information Age: the realization that any knowledge or representation—poetical to mathematical, textual to pictorial—can be expressed as a string of ones and zeroes. That perception came first for pure numbers. Around the year 1700, Gottfried Wilhelm von Leibniz, who invented calculus independently of Newton, realized that our numbering system is not unique. Decimal notation, with its digits zero through nine, seems inescapably right, but Leibniz understood that numbers can be expressed with any desired number of symbols, say five or twelve—or in the sparsest possible arrangement, the binary system (whose zero and one Leibniz associated with the void and with God).

Binary arithmetic and electronic computers were made for each other. It is difficult to build an electronic device that operates in ten different modes to represent the decimal digits, but it is simple to make a switch with just two states, on and off, interpreted equally well as one and zero. A computer is an enormous array of switches that can be turned on or off in coded patterns at high speed. The switches are set and their states communicated throughout the computer, by strings of electrical pulses and pauses that represent ones and zeroes. The more quickly the pulses come—that is, the higher the frequency of the electronic current—the more binary digits (abbreviated "bits") are sent per second. That is why, in the parlance of computer chips, a 400 megahertz computer is faster that a 200 megahertz one.

Binary arithmetic worked perfectly for the numerical calculations made by the earliest electronic computers. Calculational use remains important, but bits can also express words, sentences, and ideas; color, intensity, and imagery. To transmit the power of language, each letter of the alphabet is simply assigned its own brief string of ones and zeroes. With such coding, the old saw about a troop of monkeys pecking at keyboards to produce by chance the complete works of Shakespeare can be restated with a troop of animals that know only how to flip coins. The binary representation of letters is not new. It was in 1838 that Samuel F. B. Morse invented his two-symbol version of the alphabet, the telegraph code of dots and dashes.

A picture, too, can be put into numbers—digitized—and then into bits with the method that once offended the free vision of William Blake. He was forced to duplicate artistic prints by mechanically copying a segment at a time; the digital translation of an image breaks it into small picture elements (abbreviated "pixels"). Each tile in this electronic mosaic carries numbers that give its location in the image and that define the proportions of red, green, and blue that yield its intensity and color when they are combined. When those numbers are expressed as bits, the image joins the stream of ones and zeroes cascading through a computer or a communications system or stored on a compact disc.

In the binary world, there is no distinction among a mass of statistical data, a chapter of a novel, and a picture of a cypress tree as they are stored or transmitted. All are streams of bits that differ only in the size of each stream—although that difference is vast. Consider: Every word in a book of 200 or so pages, like this one, easily fits onto a single computer diskette, occupying about four million bits or half a megabyte (a byte is the logical unit of organization in a computer, a string of eight bits). A single image, however, in the pictorial resolution typically used on computer screens (which remains inferior to conventional ink printing) requires many times that amount; thus, a picture is worth not a thousand words but hundreds of thousands, because it requires millions of bits.

Even after electronic computers began to enter common usage, in the 1960s and 1970s, data were sent and displayed so slowly that it would have taken hours to form a single image. Now we have faster electronic methods that reduce the time to minutes or seconds. But our brains take in visual information at a billion bytes per second. To our vision-oriented intellects, the electronic transfer of bits can still seem like dipping out the sea in teaspoon-size lots. Only the enormous bandwidth of light can meet the colossal demands of manipulating images in binary form. And the small wavelengths of light allow information—whether pictures, text, or numbers—to be guided precisely down the diminutive conduits of optical fibers.

Telecommunication via optical fiber, the basis of the Information Superhighway and the Internet, represents a joint effort among light sources, detectors, and fibers. The sources are solid-state semiconductor lasers, devices far smaller than any gas laser, like the helium-neon type, and requiring much less power. When such a device is connected to a source of electrical current, electrons flow in and make photons as they drop through the semiconductor band gap. If the electrons arrive in a stream of pulses that represents binary information, the laser blinks in the same tempo, sending pulses of light over the optical fiber network. When the light reaches its destination, its photons are absorbed by a tiny detector also made of a semiconductor. It releases electrons in the same rhythm to re-create the original electronic stream of bits.

The fiber represents a deeply refined art of glassmaking. It is drawn into a thin strand, typically 100 micrometers across, by extruding the molten tip of a glass rod. In service, light races down the fiber without shining through its sides, because the fiber is made with a central core surrounded by a sheath of slightly lower refractive index. Even if a photon does not run true down the center, the same total internal reflection that makes a diamond sparkle brings the photon back toward the core, like a billiard ball bouncing between the cushioned sides of the table.[1]

The glass is made of the same silicon dioxide—sand—used in ancient glass, but it has been greatly purified. With only a few foreign atoms for every billion molecules of silicon dioxide, it is the most transparent material ever made. If one hundred photons were to enter a monstrous mile-thick plate of that glass, only six would be absorbed before reaching the opposite face. Even so, the intensity of the light would diminish below usable levels as it traversed long stretches of optical fiber, such as the four-thousand-mile telephone link across Canada. To prevent that, the light is refreshed by a "repeater," which absorbs its weakened pulses and re-emits them in the same rhythm but at renewed strength. In 1988, strands of optical fiber were laid beneath the sea for telephone service, the eighth transatlantic cable between the United States and Europe, with repeating amplifiers spaced on average thirty miles apart.

It would be a pretty scene if the vast network of optical fiber glowed with light—a lovely conceit to imagine under city streets or beneath the sea—but although there is light, there is no glow. The network uses invisible infrared light, not the visible sort, because optical fiber is most transparent at the wavelengths between 1.3 and 1.6 micrometers. Just as the addition of selenium to cadmium yellow changes the color by changing the band gap, certain combinations of the elements indium, gallium, arsenic, and phosphorus make a semiconductor laser with exactly the best wavelength for fiber optic networks. This delicate tuning, in contrast to the fixed wavelengths of gas lasers, is another reason to use semiconductor lasers.

In theory, light can be transmitted through optical fiber with a bandwidth of thousands of gigahertz. Actual values achievable today are less, but they are still many times greater than electronic bandwidths. In telephone service, this translates into enormous capacity to carry conversations. The transatlantic optical fiber transmits 300 megabits of data each second, sufficient to carry tens of thousands of conversations at once. Some traffic flows over fiber from one telephone central office to another at nearly two billion bits per second. This rate can deliver each frame of a film fast enough to form an unflickering motion picture at the receiver. The completed Information Superhighway will operate in this gigabit range, but that achievement awaits one last step. Each home is still linked to the network of glass fiber through the copper wire connected to its telephone. When that bottleneck is replaced by optical fiber, *Casablanca, Days of Heaven,* or *Star Wars,* along with telephone service and every other form of communication, can pour down a thin glass strand and into our homes.

Fast-moving photons may also supplant flowing electrons in computers, perhaps even without a material medium. In 1990, Alan Huang of Bell Laboratories built the first computer that used optical elements, although it also used conventional electronics and performed only one specific task. In 1993, a prototype general-purpose computer that employed light nearly exclusively was demonstrated at the University of Colorado. Its radical design proves the principle of light-based computing, using sources, detectors, and optical fibers developed for telecommunications. Instead of data represented by electrons stored in silicon chips, its bits are pulses of light that continually circulate through three miles of optical fiber. By accurately timing the pulses over carefully measured distances, the system tracks each binary one and zero—a twelve-foot shaft of light and a dark patch, respectively. Each photonic bit is routed as necessary by optical switches that can direct it to either of two paths.[2]

This ghostly switchyard for bits, operating at light speed and at high frequencies, is beguiling enough, but its inventors, Harry Jordan and Vincent Heuring, are working on a more complex system, with optical switches that can select from among many paths. The directed beams of light would not use optical fiber; they would simply transmit data in open space. Such a gossamer network of light could conceivably play a role in artificial intelligence. Some researchers believe that computers can evolve into man-made brains that attain the power of living intelligence. Others, weighing the three-dimensional intricacy of the brain with its billions of neurons, doubt that this ambition can be realized. But if light can be made to carry quantities of information as it crisscrosses in space, perhaps it can simulate the brain in a way that electronic computer chips cannot.

That possibility refines an older vision that associated artificial life with intense electrical effects. In Fritz Lang's 1926 film *Metropolis,* a seductive robot surrounded by a crackling electrical halo was a disturbing presence. In 1931 Boris Karloff lurched into monstrous life on screen as Dr. Frankenstein tapped the power of a lightning bolt. Now a mere trickle of electricity activates silicon computer chips, which show some aspects of human intelligence. Perhaps the next stage will return to Frankenstein's lightning bolt, but scaled down to quiet pulses of light that carry meaning as they shuttle through a three-dimensional network. I had always thought that Commander Data, the charmingly literal-minded android in the *Star Trek: The Next Generation* series, ran on a life's blood of electricity; perhaps the truth is that if his skin is cut, he leaks neither blood nor electrons, but simply glows.

Although computers based on light, let alone artificial life, lie in the future, the tiny components of the Information Superhighway are here and now. In extreme contrast, so are gargantuan systems of light with other uses: huge lasers to ignite nuclear fusion, enormous synchrotrons to probe matter. Fusion, the process that makes the sun hot and bright, has so far defeated attempts to bring it to Earth as a clean, inexhaustible source of power. It is the inverse of fission, used in nuclear piles and atomic bombs, in which an atomic nucleus splits into shards and releases energy according to the relation $E = mc^2$. The same conversion of mass to energy dominates fusion, the basis of the hydrogen bomb, but there it comes as hydrogen nuclei (protons) join into helium nuclei. Positively charged protons fiercely repel each other and can be made to fuse only at extreme temperatures and densities. These intense conditions are natural in the cauldron of the sun, but they stretch technology to its limits when we try to create the laboratory equivalent of a star.

One approach holds the hot, charged particles within strong magnetic fields, since no tangible material could withstand the ferocity of a fusion reaction. This magnetic confinement has achieved some success, although not enough for commercial use. In a second method, laser fusion, immense lasers deliver energy to a pellet one-tenth of an inch across, filled with a few milligrams of deuterium and tritium. (These forms of hydrogen, whose nuclei contain one and two additional neutrons respectively, fuse more easily than does simple hydrogen.) The barrage of photons drives the deuterium and tritium nuclei inward toward the center of the pellet. When they are compressed to a hundred times the density of lead and heated to hundreds of millions of degrees, they are expected to ignite and fuse into helium, releasing power like a microscopic hydrogen bomb. If more power is emitted than is absorbed, the process could become a valid source of energy.

Laser fusion experiments have until recently been secret, for their scaled-down simulations of a hydrogen bomb have had implications for national defense, but their astounding lasers are well known.[3] The largest is the NOVA system at the Lawrence Livermore National Laboratory in Livermore, California, whose light comes from a medium made of glass and lies in the infrared at a wavelength of 1.06 micrometers. The light emerges from ten different arms arranged around the fuel pellet, and for a few nanoseconds it delivers more power than all the generating stations in the United States combined. The photons vaporize a metal sheath around the pellet to produce short-wavelength X rays. These make the pellet implode evenly, which is crucial for fusion.[4]

This impressive system of light may be only a precursor to a bigger one, the National Ignition Facility, proposed to be built at Livermore by the year 2001. Estimated to cost $1 billion and to be housed in a structure five hundred feet long, NIF would bring two hundred laser beams to bear in a do-or-die attempt at controlled thermonuclear power. A preliminary plan for the facility shows an array of conduits for the beams: Like radiating legs connected to the body of a sixty-foot-tall spider, they terminate in a central chamber that holds the tiny pellet of fuel. The device will not be a commercial source of power— it would take decades more to develop—but NIF would complete a satisfying loop in the history of light, for conditions inside the pellet would approximate those during the first sixty seconds after the Big Bang. This means that light, in the form of powerful laser emissions, becomes the agency that investigates the era of its own birth.[5]

While it does not re-create the early universe, a synchrotron light source impresses through its sheer size. It is descended from the cyclotron developed by physicist E. O. Lawrence in the 1930s to study elementary particles, an invention that began a line of machinery that would have culminated in the fifty-four-mile loop of the Superconducting Supercollider, now abandoned in Texas. In the synchrotron, electrons enter a tunnel formed into a horizontal ring measuring up to thousands of feet in circumference. The electrons are held in orbit around the ring by strategically placed magnets while they receive bursts of energy that spur them to nearly the speed of light. As the energetic charged particles swing around the circle, they deform the electromagnetic lines of force that tie them to the rest of the universe. The result is a cascade of radiation millions of times stronger than that emitted by any conventional source. It covers wavelengths from the X-ray region to the infrared. The light emerges in an extremely narrow beam, so its full power can be delivered to a small area, and the light pulses on and off as each group of electrons circles the ring, like a great photographic flash unit blinking on a time scale of trillionths of a second.

This mighty source of photons illuminates the properties of solid matter and of biological systems from molecules to organisms. Unlike elementary particle physics and space science, research in these areas has traditionally flourished at the small-scale, tabletop level. Now individual researchers rely on these centralized light sources, a new kind of big science that enriches the meaning of a synchrotron site. I have visited several but know best the one at which I ran experiments—the National Synchrotron Light Source (NSLS) at the Brookhaven National Laboratory, located on Long Island fifty miles east of Manhattan.

When I returned there not long ago, I drove, but even from the air it is easy to pick out the synchrotron because its form so clearly follows its function. Its circular building echoes the huge ring around which electrons race to make X rays. Linked to that ring is a smaller but still enormous doughnut for ultraviolet and infrared light, where I ran my experiment. Its hangarlike enclosure has an industrial look—exposed girders, a crane to lift massive machinery, pitiless heavy-duty lighting. Mounds of research equipment fill the area around the ring with obscure shapes in stainless steel, festoons of electrical cabling, and displays of flashing digits that embody data on the fly. The confusion is typical of a working laboratory, but the essential design of the synchrotron imposes an underlying order. The apparatus is arranged in clusters, each sitting at the end of a pipe—a beamline—from which photons emerge, like a farm field watered by an irrigation channel. Each beamline supports experiments sponsored by a given institution or consortium. The result is one of the world's denser concentrations of scientific effort, with nearly a hundred research stations at the X-ray and ultraviolet rings.

In this surreal environment, it is comforting to see that each experiment is based in a human-scale lair fashioned from desks, chairs, computers, and racks of electronic gear. These dens surround the synchrotron like huts around a campfire, but the inhabitants are rarely visible among the thickets of equipment. Only traces appear—empty soft-drink cans, a chessboard and pieces atop one cabinet, a toy pink flamingo peering down from another, a sign that pleads PLEASE DON'T FEED THE SCIENTISTS. Any scientist sighted through the hanging electrical vines is likely to be a younger member of the species. Graduate students and research associates just past their doctorate do much of the day-to-day taking of data.

Except when it must be closed for maintenance or modification, the synchrotron runs hard, steadily pumping out photons to drive down their unit cost. Data may come at any hour of day or night, and it is never easy to get all the delicate equipment on a given beamline operating well at the same time. The motivation is strong to keep taking data as long as everything works. This accounts for the all-night atmosphere of crumpled cans and empty coffee cups, which I could see especially well when I carried out research at NSLS, as my experimental station was located above the ring.

Apart from the demanding hours, the research floor is a difficult place to work, with its harsh lighting and ceaseless background noise. It is not that photons pop out of orbiting electrons with cracks of sound, like tiny lightning bolts with minute thunder. Photons are noiseless, even as they are created; lightning makes thunder only as it affects the air through which it flashes. But the synchrotron and the beamlines are evacuated to keep air molecules from deflecting electrons and photons. The large mechanical pumps that maintain this emptiness produce the unpleasant noise. And always there is the knowledge that uncontrolled synchrotron radiation can be dangerous. Horns blare and lights flash every few hours when new electrons are about to be pumped into the ring, because that carries a danger of escaping radiation. Everyone leaves the research floor before any such fill.

The synchrotron is also a wonderful place to work. Its demanding aspects lend urgency and mystique to this large enterprise of light, where research is at the cutting edge. The cheek-by-jowl conditions bring a remarkable cross-fertilization among scientists whose only common interest may be what light can do. When I ran there, I continually encountered friends and colleagues doing a variety of science within a few steps of my own station. Such interaction is a fruitful aspect of this particular brand of big science. To some extent, it eases old images of science as a lonely enterprise.

Among the hundred-odd beamlines, uses across fields of science come thick and fast. Much of the work is fundamental, like the effort in which I was involved, a study of one of the fascinating materials called superconductors. These have a seemingly magical property: When cooled to a certain temperature, they lose all electrical resistance, carrying current without losing any of its energy as heat. This happens in certain solids whose quantum rules permit cold electrons to march in lockstep like trained troops, rather than in chaotic motion like a fleeing crowd. The effect has been known since 1911, but its perfect efficiency never had a dream of commercial application because the necessary temperatures were impractically low, near absolute zero. In 1987, however, the physicists Johann George Bednorz and Karl Alex Müller, of the IBM Research Laboratory in Zurich, made an astonishing discovery that gave them a Nobel Prize. They found a new class of "high-temperature" superconductors that lost resistance at much warmer temperatures, still far below ordinary cold but that could be easily reached by refrigeration. The discovery set off a world-wide scientific frenzy to understand and use the materials.

Our measurement was made to see if one of these complicated compounds, called YBCO (yttrium barium copper oxide), obeyed the prevalent theory, which predicts that a superconductor has a gap in energy somewhat like the band gap of a semiconductor. If the gap existed in YBCO, it would show its presence by absorbing certain infrared photons, as cadmium yellow absorbs certain colors. But there was a problem: Our sample was a thick slab of material that transmitted little light. Only the synchrotron provided enough infrared power to penetrate the sample, which enabled us to carry out the first such measurement. Our data indeed seemed to show the predicted absorption but with unexpected complications that made the interpretation uncertain. Despite this and many other efforts, the high-temperature superconductors remain puzzling. Infrared data, however, give important clues, and our results could not have been obtained without the high power of the synchrotron.[6]

The fact that synchrotron power emerges in a narrow beam makes possible other unique research, such as the X-ray analysis carried out on a strand of the element bismuth less than one-hundredth the thickness of a human hair. Bismuth is interesting in its own right, for it is akin to both a semiconductor and a metal, but the value of this measurement was in establishing that the tiny filament was of crystalline form, making it probably the smallest crystal ever observed.[7] Novel man-made materials, especially biological ones, are often made first in extremely small lots, which can be examined by the fine beam of synchrotron light.

Other work at NSLS involves surface science, the study of how atoms and electrons behave at interfaces, such as that between the semiconductor silicon and air or vacuum. The behavior deep inside a crystal is understood through its unrelenting atomic geometry, but the regular array of atoms stops abruptly at the surface. The boundary region is difficult to describe, which raises fundamental questions about surface behavior. The answers are significant for the semiconductor industry, as silicon chips must be made with pure surfaces before they can be turned into devices, and for industrial processes based on chemical reactions, such as the refining of oil and the manufacture of plastics, which depend on catalysts—compounds that accelerate chemical reactions without themselves changing. Many catalytic reactions occur best at interfaces; for instance, the catalytic converters used in automobiles contain platinum or palladium arranged with maximum surface area to clean polluting chemicals from the exhaust.[8]

These industrial needs benefit from fundamental studies of surfaces. Whenever atoms of one type attach themselves to a surface of a different sort—say, oxygen on silicon—they vibrate at frequencies that can be analyzed to determine their arrangement and linkages. The oscillations can be examined by infrared light, but at wavelengths, as it happens, where adequate sources have been in short supply, until infrared synchrotron light became available. Other synchrotron studies use ultraviolet photons in a kind of photoelectric effect that drives electrons out of atoms at the surface, to determine the electronic energies. X rays from the synchrotron provide short wavelengths to resolve the geometric arrangement of surface atoms, just as they give structural details for solids.

Some efforts at NSLS are explicitly devoted to industrial technology. One is X-ray lithography, a means to pack silicon chips more densely with the intricate conduits that make up electronic circuits. The smaller these channels, the more devices can be crammed onto a piece of silicon. That march to smaller features has steadily increased the capacity of random access memory (RAM) chips for computers. But it is limited by the wave nature of light, which enters through photolithography, the technique that puts the pattern of channels onto the silicon. A wafer of silicon is coated with a light-sensitive chemical called photoresist. Light shines through a stencil or "mask" of the desired pattern, casting its image on the wafer. The exposed photoresist (for some types, the unexposed material) is easily removed, and the remaining chemical defines the pattern for further processing.

No matter how narrow the channels cut into the mask, the width projected onto the silicon is affected by the illuminating wavelength.

When light passes through an opening, it diffracts—that is, it spreads beyond the limits of the opening. The smaller the wavelength, the more photonlike the light and the less the diffractive effect. With visible light, the features on a chip cannot be made smaller than half a micrometer across. That is only a fraction of a human hair, but as has been demonstrated at NSLS, short-wavelength X rays reduce the scale tenfold. This translates into great gains in the density of electronic devices, and there are plans to standardize X-ray lithography for the semiconductor industry.

The patterns cut into silicon are complex, but biological patterns are more so. Synchrotron light explores them in novel ways. Molecular biologists seek to understand the structure of molecules such as proteins, which set their biological functions. X-ray and ultraviolet analysis can determine the positions of the atoms and how they are linked, but the data are not easy to come by for these large, intricate molecules. In pre-synchrotron times, X-ray data could not completely explain the structure of DNA, although they were the main clue that in 1953 led James Watson and Francis Crick to the double helix. Even later, in 1965, thousands of conventional X-ray images had to be assembled to give the first fully determined structure of an enzyme, one of the biochemical catalysts that accelerates the processes of life.

Such laborious analysis is speeded by the high power of the synchrotron, which reduces the time needed to obtain an X-ray image. And the harm X rays do to biological systems is minimized when the radiation comes in the short bursts inherent in the synchrotron. Compared to conventional methods, more data can be obtained before the sample deteriorates. This feature enters into a new research instrument, the X-ray microscope, which sees more finely than does a conventional visible-light microscope, for the same reason that X rays make finer lithographic patterns than does visible light. Because of diffraction, no detail smaller than the wavelength of the illuminating light can be seen through a microscope. With visible light, the limit is half a micrometer, or 500 nanometers. The intense but not too destructive short-wavelength X rays at NSLS are the heart of a microscope that has discerned features measuring 60 nanometers across in chromosomes, cells, and bone cartilage.

And the pulsed light of the synchrotron gives new views of biological behavior. It is molecular structure changing in time that leads to biological function, as the change in shape of rhodopsin—its atomic gate swinging shut in response to a photon—begins the act of interpreting light by the brain. Each rapid blink of synchrotron light can "freeze" a different configuration of a molecule, like a frame of an X-ray or ultraviolet film. The changes in rhodopsin, and in the amino acids that comprise a protein, have been studied in this way. Ultraviolet synchrotron light has also examined the shape of a protein occurring in the bacteria that causes Lyme disease, with its severe effects on heart, nervous system, and joints. The results elucidate how the body's immune system recognizes foreign organisms, a fundamental question whose answer may also aid in treating the disease.

Synchrotron light is more than a biological research tool, for X rays have illuminated the workings of the human body for a century. In that tradition, clinical diagnostic medicine is also carried out at NSLS. This is the project in coronary angiography, which uses synchrotron light to form images of the coronary arteries. These narrow conduits, no more than one-eighth of an inch across, carry oxygenated blood to the heart muscle itself, with serious consequences if they become choked with fatty plaque. In the standard method of examination, a liquid that absorbs X rays is injected directly into the coronary arteries, through a long narrow tube called a catheter. This enhances a conventional X-ray image sufficiently to show the condition of the arteries, but the insertion of a catheter into a coronary artery has its hazards. The risk is lower if the high-contrast liquid is injected into a vein; then, however, it is diluted by the time it reaches the coronary arteries, which degrades the X-ray image.[9]

With X rays from the synchrotron, physicians can use the safer injection into a vein. The high power and the choice of wavelengths combine to give a clear image, with the patient receiving no more X radiation than in the conventional method. Since 1990, patients have come with their physicians to NSLS. They are examined in a facility designed to minimize the inevitable reaction to this unorthodox approach: The treatment room resides at the end of a beamline, but one that has been separated from the main floor. The patient enters it without seeing the enormous machine whose light will soon penetrate his body. Still, the room is not especially comforting, but it is no worse than other medical technology many of us have confronted. I myself would far rather occupy its patient's chair than enter again the tight tunnel of a massive MRI machine, in which claustrophobia can approach panic proportions.

That chair at NSLS, resembling an old-fashioned barber chair, rests in line with the X-ray beam from the synchrotron. There the patient sits after the high-contrast liquid has been injected into a vein. The members of the control and observation team work in an adjacent room, viewing the scene through glass that blocks them from the X radiation. They move the chair by remote control to briefly align the patient's heart with the beam. In a few minutes, the image appears on a computer monitor, ready to be examined. Although the technique is highly promising, there is one worrisome aspect—the cost to build a single-purpose synchrotron devoted to clinical medicine. Improvements, such as efficient magnets to control the electrons, can reduce size and cost, but even a vastly scaled-down synchrotron is in the ten-million-dollar range.

During my visit, I spoke with Gwyn Williams of NSLS, who had provided the synchrotron expertise and equipment for our joint superconductor measurement. This mustachioed, British-born scientist was dressed casually, like most working researchers, in jeans and running shoes. His devotion to the machine never flags, and he is always ready to talk about it, with such intense focus that he usually answered my question before I finished asking. He emphasized how young synchrotron science really is. NSLS has been running for over a decade, but it "takes ten years to get a synchrotron fully tuned up," he said. Improvements in this complex machine are still coming. Recently, for instance, it was noted that the beam of electrons wandered slightly in space as it dashed through the synchrotron tunnel. New controls have reduced this drifting, resulting in more stable light. Such improvements, along with novel approaches, are appearing in other synchrotron designs. Synchrotron laboratories, in fact, are springing up around the world—about forty, ranging in completion from proposed to operational, in fifteen countries. The newest in the United States include the Advanced Photon Source, whose ring two-thirds of a mile in circumference is being constructed at the Argonne National Laboratory in Illinois, and the Advanced Light Source at the Lawrence Berkeley Laboratory in California, located in the building that housed a forerunner, an early cyclotron.[10]

Powerful X-ray light, from synchrotrons and from ordinary sources, is essential for medical diagnosis. But "small" light—weak photons from low-power sources—is emerging as a new tool that cannot harm the body. One technique takes advantage of the fact that flesh transmits visible and infrared light, an effect you have seen if you ever watched a flashlight shine pinkly through your hand. The translucency was used in the nineteenth century to detect tumors, which sometimes cast shadows in transmitted light. Now the method, called transillumination or optical tomography, draws on photonic technology to give detailed images.

In a configuration now undergoing trials, David Benaron and David Stevenson, of Stanford University, use infrared light from semiconductor lasers sent through optical fibers. The light is attenuated up to a hundred million times as it passes through the subject, but it can still be detected. It is analyzed by computer to take into account the possible path of each photon as it is absorbed, scattered, or transmitted by the biological structures it encounters. The computer selects the earliest-arriving photons, which have suffered the fewest multiple interactions and hence carry the clearest image. In one demonstration, light that had passed through a dead laboratory rat unmistakably showed its heart, liver, intestines, and other internal features. Although the resolution of the method is not yet extremely high, its noninvasive nature is a valuable asset. Benaron and Stevenson, as well as other researchers, are working on variants to monitor oxygen levels in the brains of infants and to diagnose breast cancer as an alternative to X-ray mammography.[11]

Weak light is also beginning to map the workings of the living brain. Researchers at the University of Washington have examined the cerebral cortex on the surface of the brains of patients undergoing surgery for epilepsy. As part of that procedure, portions of their skulls were removed under local anesthetic. The investigators shone infrared light on the cortex and detected the reflected light. The reflection from certain areas changed when the brain was stimulated with electrodes and when a patient exerted his facilities for language by naming objects, probably because of altered blood flow or chemical conditions at the active site. Optical determination of such critical areas would offer guidance during brain surgery. And if this technique can be made to function through the skull, as is now under study, the method may some day become a full-fledged noninvasive means of scanning the brain.[12]

Amid these satisfying medical advances, it is essential to remember that photonic technology, like any other, carries different meanings for society. Photonics has developed partly in response to military need and with military support. Laser weaponry is the obvious application, at least judging from the prevalence of destructive rays in science fiction. The Martian invaders in H. G. Wells's *War of the Worlds* devastated humanity with an invisible heat ray; in later works of the imagination, Flash Gordon and Han Solo also wielded ray guns. These fantasy weapons can be rationalized as employing laser light, whose military potential has indeed been explored for nearly four decades. In 1958 the Defense Department began to fund laser research, at the TRG Corporation. The proposal from this defense contractor was based on the patent application of Gordon Gould, the inventor of the gas laser. Gould has noted that the funding agency was "exceedingly happy about the prospect of a death ray, Buck Rogers style" and funded the proposal at a million dollars—triple the amount requested.[13]

Since then, billions of dollars have been invested in high-energy military lasers. X-ray lasers powered by nuclear reactors were meant to be part of the Strategic Defense Initiative (SDI) proposed by President Ronald Reagan in 1983. They would have been placed in orbit around the Earth, ready to destroy intercontinental ballistic missiles. X-ray lasers were not scientifically achievable, however, and this program has been terminated. Funding for large lasers has declined in the post–cold war world, and also because it has proven difficult to design portable power supplies of sufficient capacity (which makes laser hand-weapons unlikely). But one type of powerful laser needs no electricity. Its light comes from chemical reactions that can emit megawatts of infrared power. Such a chemical laser is incorporated into the battle-field weapons system called GARDIAN (General ARea Defense Integrated ANti-missile), designed to find and shoot down missiles at distances to seven miles. Another, the Alpha laser, is meant for use against intercontinental ballistic missiles.[14]

Whatever the future of laser weaponry, photonics is imbedded in the so-called C³I functions of warfare—command, control, communications, intelligence. In the nineteenth century, military photonic communication meant the heliograph, a mirror that reflected a beam of sunlight to an observer miles away. The beam was blinked on or off by tilting the mirror to send a message in Morse code. The method was valued for its speed and its secrecy—information was confined to a narrow beam rather than widely broadcast. The same military virtues reside in communication by optical fiber; it is more difficult to tap than electronic communication and emits no radiation that would reveal its presence. Optical fiber gives a secure means to control a battlefield missile, without burdening it with guidance electronics. The flexible strand pays out behind the weapon, forming a conduit that carries data to and from an electronic or human controller at the launch site. Some military aircraft now "fly by light," with optical fibers linking their internal electronics. The fiber is lighter than copper wire and is less affected by the huge electromagnetic pulse generated by nuclear explosions.

Photonics also provides intelligence, reconnaissance, and target acquisition. Those abilities depend on technology that began in World War II, when snipers used a "snooperscope," a portable infrared detector that located targets at night via their emitted heat. Modern infrared methods have wide military application, often using detectors made of the semiconductor mercury cadmium telluride (MCT). This compound, which I have studied in my laboratory, is a cousin of the artist's pigments cadmium yellow, orange, and red (cadmium sulfide with added selenium or mercury). As these pigments absorb certain colors, MCT absorbs infrared photons because it has the appropriate band gap.

Infrared methods continue to be used on the battlefield by infantry-men, and they also direct missiles to the hot engines or exhausts of armored vehicles and aircraft, an important element in the Gulf War of 1991. Infrared surveillance also gives a global perspective, through detectors mounted in aircraft or earth-orbiting artificial satellites. The United States has long placed infrared and visible light sensors in satel-lites to detect launches of intercontinental missiles. When such a rocket first climbs into the air, it leaves a long trail of hot exhaust gas. The resulting infrared radiation stands out against the cooler back-ground beneath, even from satellites placed high enough to observe broad expanses of the Earth. This wide observation was planned as the essential first layer of SDI.

Warfare requires a directed use of photonics that seems to be the antithesis of its free use to enhance life, but the truth is that the poles are often linked through dual civilian and military application. The optical fiber that brings people together over the Internet also guides a missile to its target; the infrared imaging that senses the enemy also attacks Lyme disease, for the ticks that carry the illness infest deer, whose numbers can be counted even in thick woods by tracking their heat.[15] And the technology of military surveillance from satellites and aircraft is a powerful tool to examine our world. That is the photonic application of remote sensing, which began in 1858 with the first aerial photograph, snapped from a balloon hovering over Paris.

In modern remote sensing, detectors respond to visible and infrared light from the sun after it has been reflected, or absorbed and reradiated, by land, sea, or atmosphere, and they note infrared radiation from the internal warmth of the Earth. Properly interpreted, photographs and spectral data gathered in this way give an astonishing range of information. Remote sensing helps predict weather with data from a network of meteorological satellites; it examines the condition of the environment and of crops; it aids in the search for mineral resources by identifying features of soil and geological structure; and it examines the temperatures of the ocean as well as viewing its waves and currents. Satellite data gave detailed information about the great Midwestern floods of 1993. The images spanned the affected states with pixels thirty-one miles on a side, showing the degree of ground saturation and flooding to that resolution.[16]

Civilian and military uses of photonics reflect different values that guide its application. A similar collision of values appears when photonics interacts with visual art. The meaning of an original work of art must be rethought when its image can be delivered on demand from a CD-ROM or over the Internet. Many are delighted and informed by this wide access. Art can be newly appreciated on a CD-ROM, whose immense storage encourages comparisons and carries details of artists' lives and artistic movements. But many are dismayed at art quantified, which makes it possible to paint a digital mustache on the Mona Lisa.[17] That accelerates a trend which already puts Van Gogh's works on mass-produced neckties and may end in an uncanny ability to clone artwork. In an experiment by the Hughes Aircraft Company and the J. Paul Getty Museum, a laser scanned a figurine and stored its contours in a computer as a set of instructions to a second laser. Its ultraviolet light solidified a liquid plastic until, layer by layer, it built up a three-dimensional replica.[18]

Yet photonics also deepens the meaning of "original art," for the technology authenticates paintings and shows how they are created.

With infrared, X-ray, and ultraviolet light, experts peer through the top layers of pigment to earlier strata. The subsurface images may reveal pentimenti, or changes the artist made as he painted (after the Italian *pentirsi,* "to change one's mind"). These methods have probed works by Titian and Rembrandt, among others, and enter into new attributions. In 1992, the work *Madonna with the Pinks,* thought for a century to be a copy, was identified as an original Raphael: X-ray and infrared examination at the National Gallery in London showed a characteristic series of underlying circular forms that blend into Raphael's softly rounded look for the Madonna.[19] Analytical light also affects views of modern art. When the Picasso scholar William Rubin examined infrared photographs of *The Pipes of Pan,* he found that Picasso had eliminated two of an original four figures as he painted. Rubin identifies one deleted figure as a Venus based on the socialite Sara Murphy, who with her husband Gerald befriended Picasso, F. Scott Fitzgerald, and other luminaries of the 1920s. Her erasure came when she rejected Picasso's romantic attachment, which had led him to portray her in other works.[20]

Photonic technology exalts art and diminishes it, enhances both warfare and medicine. It is right to apply moral categories to these outcomes but not to photonics itself, which is neither good nor bad; only rich and complex, like the wave-particle duality of light. We cannot say that light is solely wave or solely photon. We know only that how we choose to look at it—what experiment we do—reveals one face or the other. With technology, it is only human choice that uncovers a fair face or a dark one, or both at once.

One of the fairest uses of photonics can be envisioned as arising out of military applications. Think of a surveillance satellite floating in orbit above the Earth's surface, its photonic eyes narrowly focused to seek only the plumes of ascending missiles. But that vision can be broadened to examine the whole sweep of the planet beneath. And it can be deepened if we turn the photonic eyes outward, toward sun, moon, and other planets; toward the gulf beyond the solar system that leads to the heart of our galaxy; toward vaster gulfs opening up to further galaxies.

This search for cosmic light is the latest form of an ancient fascination. Its first tool was the naked eye; then the glass lenses and mirrors that Galileo and Newton formed into telescopes. Enormous ground-based telescopes followed, some devoted to invisible light. Finally came today's space-going systems whose names and acronyms—Hubble, COBE, EUVE—conceal miracles of astronomical science that probe light.

Cosmic light has always drawn the mind and spirit. It has always spoken to the mysteries of creation, until we seem on the verge of grasping even the Big Bang through the record it has written in light. Now our observation and interpretation of cosmic light bring together all our knowledge of the universe: its relativistic behavior; the meaning of the last great elementary force, gravity; the quantum world of photons, and of the technology that detects them. Cosmic light has become universal light, which combines each of its nearly infinite aspects.

8

Universal Light

[G]reater telescopes will be put into operation. And slowly, as the darkness recedes, the universe will loom forth.

—Edwin Hubble, extragalactic astronomer

In the 1950s, before we entered interplanetary space and walked on the moon, the artist Chesley Bonestell carved out a particular niche for himself. He painted scenes set throughout the solar system and into the farther universe, envisioning as realistically as possible the lunar surface, the planet Mercury, the rings of Saturn, and the distant star Mira. His majestic images seemed to transport the viewer to alien worlds. They delighted popular audiences and science-fiction readers, yet those same accurate images illustrated the seminal book *Conquest of Space,* written by Willy Ley in 1950, which seriously examined goals in space exploration. Views of the cosmos, whether of nearby planets or remote galaxies, represent one beating pulse of astronomy: the desire to experience the universe directly, to see it through our own eyes. The scientific care in Bonestell's art represents another ancient thread—analysis of cosmic light to understand the universe.

That same mixture of vision and analysis was apparent when

humanity watched fragment after fragment of Comet Shoemaker-Levy rain onto the planet Jupiter in the summer of 1994. The spectacle was examined by telescopes, analyzed by spectroscopic instruments, and converted into digital images and data for scientific study. The images were also sent over the Internet to hordes of avid viewers at their computer terminals. These enormous flourishes of light—which could just as well have come from the sun or the stunning Andromeda Galaxy two million light years away—became technological light squeezed into fine tendrils of optical fiber, and then perceptual light glowing from monitor screen to be received by human eyes.

Although we now convey cosmic sights along a web of light, we still feel the shock of reality under the naked night sky and strive to experience what lies beyond our world. We have seen the moon from its surface, have sent our robot representative to examine the soil of Mars. Unmanned spacecraft present us with closeup images of distant Jupiter and Uranus, but it may be a long time before we walk upon the outer planets, travel to a star, or spiral into the depths of a black hole. Until then, we glean what we can from cosmic light. We take away visceral awe and a response to natural beauty, along with the cooler reaction of the analyzing intellect, as we contemplate planets, stars, and galaxies.

Apart from the sun and the moon, all that humanity first saw in the sky was motes of light without internal detail. Some shone brightly, others were barely visible; some were white, others carried color; some wheeled slowly across the skies, others changed appearance on a nightly basis. Those observations differentiated between the first two classes of astronomical objects: distant unmoving stars glowing with their own energy and nearer planets that reflected light without making it. Later came a bewildering array of other entities—supernovas and nebulas, galaxies and quasars. Some were brilliant beyond belief, showing the power of their internal processes. Others were black beyond all previous meaning or were invisible, yet these black holes and dark matter also show their presence as they affect light.

The cycles of the sun and stars must have conveyed a sense of order to early watchers. Human eyes and imaginations imposed spatial order as well, by picking out constellations of stars that defined stellar neighborhoods. The oldest known catalog of the skies was compiled around 150 A.D. by the same Ptolemy of Alexandria who presented a geocentric theory of the solar system. Even limited to what his naked eye could see from his Mediterranean locale, it shows more than one thousand stars grouped into forty-eight constellations. These stellar groups—such as Andromeda, which contains the Andromeda Galaxy, the galaxy nearest our own—stay with us yet. Most of the eighty-eight constellations that now define the visible sky are associated with Greek mythology or earlier cultures. In modern astronomy, computers armed with mathematical coordinates guide telescopes, yet Taurus the Bull, Orion the Hunter, and other mighty, part-imagined figures in the sky remain lodged in our vision and consciousness.

The first stargazers could not resolve points of light in the sky, but they could determine their positions by naked eye. Early observatories in China and central Asia used fixed markers to guide vision along selected lines of sight, as did the huge rough monoliths of Stonehenge. The last observatory before the arrival of the telescope was built in 1576 by Tycho Brahe, a Danish aristocrat turned astronomer, with an unenviable distinction: His nose had been cut off in a duel over a mathematical problem. He replaced the missing organ with a painted silver one, but his disposition, which combined the "irritability of a genius with the haughtiness of a nobleman," had not benefited.[1]

Brahe was, however, a careful observer, measuring the positions of stars and planets with quadrants and sextants. These carried one-quarter or one-sixth of a circle marked in degrees, as on a protractor. A movable arm was aimed at an astronomical object, and its angle was read from the calibrated arc. Using devices up to nineteen feet long, Brahe located planets to an accuracy equivalent to the width of a dime as seen four city blocks away. In 1609 Johannes Kepler used these excellent data to determine that the orbit of Mars is elliptical. This stroke destroyed the idea of the "perfect" circular orbit as a central principle in the universe. Eventually it confirmed Nicholas Copernicus's placing the sun at the center of the solar system, which he had done early in the sixteenth century.

There was more to Brahe's observations than sheer measurement of position. In 1572 he saw a great and rare spectacle: a new star in the constellation Cassiopeia, so brilliant that it shone during the day. His new star was a supernova, an exploding star that glows at a billion times its normal level. Humanity has seen only a few other supernovas from stars within the Milky Way. One in the year 1054 A.D. left behind as debris the glowing mass called the Crab Nebula, and one in 1604 was observed by Kepler and by Galileo. No one then knew what caused these great doings of light or, for that matter, the daily glow of the sun.

Patterns of light in the heavens, spiced with the occasional spectacular event, were a beginning, but the light had to be brought more definitively to Earth. Galileo used the telescope immediately after its invention to enhance images carried in from space, a first great step in observational astronomy; later William Herschel, Joseph Fraunhofer, and others analyzed light to initiate the science of astrophysics. Further advances gave better lenses and mirrors to form images, new instruments to separate the wavelengths of light, and photographic emulsions and solid-state detectors to detect and record it. These have culminated in great telescopes like those on Mount Palomar in California and the Mauna Kea volcano in Hawaii or floating in space like the Hubble instrument. Equipped with sensitive detectors, they analyze visible light nearly one photon at a time, along with X rays, ultraviolet light, and infrared light. It is the totality of cosmic light, visible and invisible, that relates the long theme of cosmic history and the shorter tales of planets and stars.

There was a great deal to discover even before invisible light was known, as Galileo found in 1610 when he turned a telescope to the heavens. Galileo was born the year Michelangelo died, and he died the year Newton was born—a fitting range for a life that combined the soul of the Renaissance with the spirit of modern science. Unlike Newton, Galileo was no cloistered researcher. He had artistic and musical talent, and he wrote with persuasive flair. He loved to debate his ideas, using devastating sarcasm against his opponents. He fathered three children and reveled in the pleasures of society and of wine, which he called "light held together by moisture."[2]

Galileo might have lived a plainer life had he followed his early inclinations, for at age fourteen he entered a monastery. But his father objected, and instead Galileo began medical study at the University of Pisa. His true life in science began, however, when he turned to mathematics. As a professor of mathematics at Pisa, he studied the laws of moving bodies. It was from the Leaning Tower that he supposedly dropped objects to disprove Aristotle's idea that the speed of a falling object is proportional to its weight. When the blazing supernova of 1604 caught his eye, he analyzed its light and concluded that it originated at a great distance.

That supernova may have whetted Galileo's appetite for cosmic viewing; in 1609, when he heard of the invention of the telescope, he put aside his other work to make his own, with a convex objective lens (that nearer the object) and a concave lens in its eyepiece. Galileo showed this device to the Venetian Senate. The senators were impressed by its value for a seagoing nation, and they offered him lifetime tenure and a substantial salary at the University of Padua, but he chose instead a position in Florence as court mathematician to the Grand Duke Cosimo II de Medici.

That first telescope magnified ninefold, and the one Galileo aimed upward in January 1610 magnified by the respectable factor of thirty. Magnification alone, however, is of no value if the telescope brings light to a blurred spot rather than a fine point. One problem is chromatic aberration, in which a lens focuses red and blue rays to different places. And if the glass is not accurately shaped, if its surface is rough, or if it contains bubbles, the image is further degraded. A good objective lens must be made with care. That care is apparent in the lens, two inches across, through which Galileo first saw four moons of Jupiter. He presented the lens to his patron the Grand Duke, and it survives today in an ivory frame. Recent testing shows it to be of good optical quality even in modern terms. Galileo achieved this by sorting among quantities of lenses he bought from spectacle makers, selecting only three of three hundred as meeting his standards.³

Galileo's telescopic observations, published in March 1610, examined planets, moons, the sun, and stars. In addition to the Jovian satellites, he saw the rugged mountains of our own moon, resolved the blur of the Milky Way into separate stars, observed "ears" on Saturn which his telescope could not quite discern as rings, and saw moving spots on the sun, which he took correctly as evidence that it rotated. A little later, he noted that Venus had phases similar to those of our moon. This persuaded him that Copernicus was right to put the sun at the center of the solar system, for only that geometric arrangement would illuminate Venus as he saw it.

In 1632 Galileo presented his conviction that the earth circles the sun in his *Dialogue Concerning the Two Chief World Systems,* a document that upset theological suppositions. The book was banned; Galileo was called before the Inquisition and then was placed under house arrest. (If he truly said, "And yet, it moves," as heroic legend has it, he did so during that period of arrest and not during his ecclesiastical trial.) Galileo may have paid another price for his stargazing: He complained of damaged sight after gazing at the sun, and although there is no direct evidence of long-term effects, perhaps something lingered, for he died blind.

Martyrdom or no, Galileo's telescopes inspired larger devices and the observatories to house them. The Paris Observatory was completed in 1672, and the Royal Observatory at Greenwich, England, was designed in 1675 by Christopher Wren. The quest for knowledge may have inspired these early forms of big science, as it did the Superconducting Supercollider in Texas, but like that abandoned modern project, they were also expected to yield other benefits. The sumptuous Paris Observatory reflected well on Louis XIV. The seafaring British built Greenwich Observatory for "the finding out of the longitude of places for perfecting navigation," as Charles II wrote when he established it.⁴

The larger telescopes became, the more effectively they examined the skies. When Galileo pointed his handheld instrument at the star cluster called the Pleiades, he saw only five stars at once. By comparison, a modern telescope, with an aperture forty-eight inches across, embraces one hundred fifty members of the cluster and twenty thousand dim stars in the background.[5] A larger aperture gathers more light, and reduces undesirable diffractive effects. As light waves bend around the edge of a barrier they create a set of bright and dark bands rather than a knife-edged shadow. In astronomy, diffraction at the rim of a telescope tube surrounds any image, such as that of a star, with diffuse rings of light that obscure fine detail. The effect is less pronounced with a large aperture, for the path of the light is more nearly straight relative to a large opening.

Large apertures need large optical elements, and this is easier to achieve with mirrors than with lenses. Newton designed and made the first reflecting telescope in 1669, using a curved mirror to magnify far objects. A reflecting telescope offers great advantages over a lens-bearing refracting telescope; it eliminates chromatic aberration, Newton's original motivation. Each mirror needs only one optically perfect surface and reflects almost all the light it receives, whereas each lens requires two excellent surfaces and absorbs light if it is large and thick, dulling the image.

For these reasons, large reflecting telescopes began to appear in the eighteenth century. William Herschel, born in Hanover in 1738, made and used the best reflectors of his time. His father, a military musician, passed to his son a love of astronomy that came to dominate Sir William's life. While working as a church organist in the English spa city of Bath, he built telescopes with his sister Caroline and began to survey the stars. In 1781 he saw something that appeared as a patch of light, not a point, and it changed position overnight. Initially he thought it was a comet, but it proved to be the first planet discovered by telescope, Uranus. In honor of that discovery, King George III named Herschel his private astronomer at a salary of £200 per year. Thereafter Herschel devoted himself to astronomy. He found ingenious ways to make large metal mirrors, and in 1789 he built a reflector with a four-foot aperture. Its forty-foot-long tube was aimed at the skies with the aid of an intricate tackle of ropes and pulleys, like the rigging on a sailing ship.[6]

Herschel's telescopes were especially effective in finding the mysterious nebulas. The name means "cloud," and these faint patches of light hovering in the telescopic field had puzzled observers since the seventeenth century. Immanuel Kant speculated that they might be groups of stars like our own Milky Way galaxy. In 1781 the Frenchman Charles Messier prepared a catalog of about one hundred nebulas to avoid confusion with the comets he sought, which look similar in the telescope. His entries are denoted by his initial, so the nebula in the constellation Andromeda becomes M31. In a decade's worth of gazing, Herschel added thousands more to the list. Some seemed to contain stars; others he thought to be clouds of gas. Their true discrimination, however, required spectroscopic vision, which was to come later.

Royal support of Herschel's activities paid off handsomely, but other astronomical observation was privately funded. William Parsons, third Earl of Rosse, resembled the gentleman physicist Lord Rayleigh in his working at his family seat, Birr Castle in Ireland. In 1840 he spent £12,000 to build the biggest reflector of the time, its six-foot mirror fashioned from an alloy of copper and tin. In a contemporary photograph, the telescope tube is mounted next to an ivy-covered wall; a man and woman stand comfortably upright inside its gaping aperture. Rosse specialized in nebulas. He gave Messier's M1 the descriptive name the Crab, and he was the first to recognize that some nebulas took striking spiral shapes, which he observed initially in M51, the Whirlpool. His observations were interrupted in 1846 by the Irish potato famine. Rosse devoted his attentions and money to his tenants, for which he was well remembered. Thousands attended his funeral in 1867.[7]

The dramatic swirl of the Whirlpool nebula may have influenced Van Gogh's *Starry Night*. To me, the painting has always seemed to represent a direct unmediated view of the universe, different from what science knows. But Van Gogh may have seasoned its powerful surges of light with detailed knowledge, since several of its features resemble known astronomical objects. Most striking is the central spiral in the sky. Except that it is reversed right to left, it looks much like the Whirlpool. Van Gogh could not have seen that nebula by unaided vision, but in 1882 its compelling form appeared in *Les Etoiles,* a popular book by the French astronomer Camille Flammarion. Perhaps Van Gogh saw the shape there and transmuted it to marvelous heights.[8]

Although reflectors gave excellent results, lens-bearing telescopes were still widely used into the nineteenth century. In 1844 in Cambridge, Massachusetts, Alvan Clark and his two sons became the first Americans to make fine lenses. Clark—engraver, portrait artist, and inventor—used his talents to shape a disk of intransigent glass into a lens perfectly curved, perfectly smooth. The lens was first ground into rough shape with a steam-driven tool and then further formed by hand with fine abrasive powders. Clark would run his fingers over the glass, seeking invisible imperfections, and he used his bare thumbs for the final polish because no cloth was sufficiently soft. There is a kind of sacrifice in this slow erosion of glass by flesh, as Clark's daughter-in-law recognized when she said, "Into every superior work the martyrdom must come." His thumbs burst from the continual stress, and legend holds that raging infections in those poor damaged thumbs contributed to his death.[9]

My own university owns a century-old Clark telescope with a lens six inches across. Its brass tube needs polishing and its original eyepiece has been replaced by a newer design, but it is still perfectly usable. Along with its sturdy wooden storage box, it carries the mark of good New England craftsmanship. The Clarks' craftsmanship reached a pinnacle in 1897, when they completed a forty-inch objective lens for the Yerkes Observatory in Williams Bay, Wisconsin (endowed by Charles T. Yerkes, who built the Chicago transit system and whose life inspired Theodore Dreiser's trilogy about wealth and power: *The Financier, The Titan, The Stoic*). It is the largest, finest lens ever made and installed. After Yerkes, Alvan Clarke wanted to make a sixty-inch objective, but the forty-incher could be called the "last lens," for the tide turned decisively to mirrors.[10]

From their earliest days, the Mount Wilson and Mount Palomar Observatories, in Southern California, housed giant reflectors. Mount Wilson's 100-inch mirror was installed in 1918. It inspired the 200-inch reflector at Mount Palomar, the largest telescope in the world for three decades after its installation in 1948. These telescopes required the difficult fabrication of huge glass disks, but now reflectors use ingenious "virtual" mirrors. In the new Keck telescope in Hawaii, computers deploy thirty-six small hexagonal mirrors into a mosaic equivalent to a 394-inch reflector, which has just detected the most distant known galaxy at 14 billion light years. The Very Large Telescope planned for a site in the Chilean Andes will connect four mirrors through optical fibers into an equivalent 630-inch reflector. When it is completed, its resolution will be the best on Earth.[11]

New detectors of light also make telescopes more effective. In the 1970s, the photographic emulsions that had recorded cosmic light for a century began to be supplanted by semiconducting detectors, which convert photons to electrons. These are vastly improved descendants of the detector that Herschel used in 1800 to first sense invisible light: He moved a thermometer along a solar spectrum from a prism, to determine the heat carried by each color. When he placed the thermometer beyond the last trace of red light at one end of the spectrum, it continued to show a heightened temperature. Herschel understood that the thermometer was detecting invisible radiation, and he measured some properties of this new infrared light.[12] Now infrared light is sensed by the same semiconductors used in optical-fiber communications or those related to the pigment cadmium yellow. Visible light is sensed by detectors made of silicon operating at nearly the ultimate quantum efficiency, at which each photon releases one electron; they produce four electrons for each five impinging photons. In contrast, the best photographic emulsions require thirty to fifty photons to paint a single spot.

But in 1918, decades before these advances, the 100-inch reflector at Mount Wilson was already seeing many millions of light years into space. It showed great numbers of the mysterious nebulas, some so dim that they darkened only a few grains of silver on a photographic plate. Their nature and significance would have remained obscure without a separate historical thread in the understanding of cosmic light—its analysis by wavelength.

The power of spectral analysis became apparent in 1814 when Joseph Fraunhofer—eleventh child of a poor glazier, brilliantly self-taught in optics—examined sunlight through his new spectroscope. It split light into colors with a prism and added innovations to ensure that each separate wavelength appeared without overlapping adjacent ones. Later Fraunhofer introduced the diffraction grating, a flat surface etched with thousands of closely spaced lines. In the same selective reflection that makes the rainbow glitter of a modern compact disc, the grating separates light into its wavelengths more effectively than does a prism. These instruments showed new features superimposed on the familiar bands of color from sunlight, hundreds of narrow dark regions where the sun seemed not to shine.

What causes the dark Fraunhofer regions is that radiation from the ultrahot core of the sun is selectively absorbed by cooler stuff as it makes its way outward. Fraunhofer did not know it, but he was observing a quantum effect—photons exciting electrons between atomic energy levels. The sequence of dark lines is a fingerprint of the composition of the sun or any star, for each chemical element has its own unique set of atomic transitions. In 1868 features were found in the solar spectrum that did not correspond to any known element. That was the discovery of helium, named for the sun.

In 1863 the pioneering astrophysicist William Huggins first applied these analytical methods to light from beyond our sun. Huggins added a spectroscope to his modest refractor with its eight-inch Clark lens, which stood next to his home near London. He was among the first to photographically record astronomical images, learning to make long exposures to capture dim sources. The spectra he preserved on the monochromatic emulsions of the time illustrate light analytical abstracted from light pictorial. They show lines and bands in black, white, or gray, spaced in order of wavelength. One early spectrum, from the star Vega, carries features that represent hydrogen and clearly illustrates how Huggins could determine the composition of distant stars.[13]

Spectral analysis has drawn immense knowledge out of cosmic light. Some stars burn blue-white, others are yellow like our sun, or dull red. The wavelengths can be related to the temperature of the star, whereas the power radiated by the star—its luminosity—is related to its size. The properties of temperature and size classify a whole range of stellar types. At one extreme lies a supergiant star like Betelgeuse, which defines the right shoulder of Orion the Hunter. It is red, bright, and relatively cool at 4,900 degrees Fahrenheit. The character of its light correlates with an ethereal density less than one-millionth that of water and with enormous size: Were it to replace our Sun, it would reach the orbit of Mars. At the other end of the scale resides a white dwarf, such as Sirius B. This companion of the bright star Sirius was found in 1862 by one of Alvan Clark's sons. It is invisible to the naked eye, but spectral analysis shows it to be hot, at 48,000 degrees Fahrenheit; smaller than the Earth, with a radius under 3,000 miles; yet more massive than our sun, for it is a million times denser than water.

The sun's own place amid these extremes should encourage our humility, for it is nothing special. Its surface temperature of about 9,900 degrees Fahrenheit is mid-range; its official size designation is "dwarf" (although its mass and radius of 430,000 miles are near average for the stars in the Milky Way); and it is slightly denser than water. Its luminosity, although staggering in comparison to any earthly source of energy, is also ordinary on the stellar scale. There is comfort in that dimness, for a bright star burns up like a guttering candle. The most luminous live only a million years, whereas our sun should last for billions.

Stellar spectra also show that a star is a globe of hot hydrogen and helium with minute traces of other elements. For every million atoms

179

of hydrogen in our sun, there are only four hundred of nitrogen and five of iron. The presence of hydrogen and helium is the first clue that a star is a glowing furnace stoked by nuclear fusion. In the initial step of the synthesis, four hydrogen nuclei are consumed as they form a single helium nucleus with mass less than the starting value. The difference in mass appears as energy according to Einstein's relation $E = mc^2$. Further combinations make carbon, oxygen, and other atoms.

These processes give a framework to understand how stars live and die. Some end in bright suicide, a supernova like those that impressed Brahe and Galileo, and another seen in 1987 that lay outside our galaxy. One type of supernova comes from gravitational collapse, which can also make a black hole. The extreme temperatures inside a star drive its atoms relentlessly outward, while the immense gravity pulls them inward. The ball of hot gas is stable as long as the forces balance, but when the nuclear fuel is depleted and the stellar furnace cools, gravity dominates. The star rapidly falls in on itself, its atoms compressing until a pinch of stellar stuff weighs a million tons. As the star drives inward faster and faster toward its core, it gains enormous energy like any mass falling under gravity, which may be released in the vast explosion of a supernova. What is left over from the original star may become a black hole, an incredibly dense concentration of matter that produces gravity so fierce that nothing—not matter, not even light—can escape.

The colors of the planets, such as the red of Mars that even the naked eye sees, also give clues to their composition. A planet reflects sunlight that has been modified by selective absorption due to molecules in its atmosphere or on its surface. The spectrum of light reflected from Mars is like that of iron oxide, suggesting its surface composition; the blue-green of Uranus originates from methane gas in its atmosphere. Similar analysis of the light from the impacts of Comet Shoemaker-Levy on Jupiter shows traces of ammonia and other compounds that reflect the composition of the planet's atmosphere or of the comet.

Most important, analysis of light definitively shows that some nebulas are assemblies of stars, as Kant had speculated, and some are clouds of gas. In 1864 William Huggins aimed his spectroscope at the nebula in the constellation Draco. He saw a sharp glowing line of color, the spectrum of a hot gas, whereas Andromeda and other nebulas gave continuous bands of color interlaced with dark lines, the spectrum characteristic of a star. Some gaseous nebulas are spectacular, such as M42, which is visible without a telescope in Orion's sword. But the star-bearing nebulas would turn out to have the greater cosmic significance. Sixty years after Huggins recorded their spectra, Edwin Hubble would find that they lay at enormous distances and constituted the biggest chips in the cosmic game—galaxies external to ours.

Huggins gave additional insight into cosmic light when he used his modest home telescope to carry out redshift analysis. The method extended work carried out in 1842 by the Austrian physicist Christian Doppler, who had shown that the nature of a sound changes if its source is moving, as when an automobile races past you with its horn blaring. As the vehicle approaches, draws even, and recedes into the distance, you hear the horn note shift from a high pitch to a low one. When the horn is stationary, it emits sound at a given wavelength, the fixed distance between two successive wave crests. When it moves, each crest is released at a different place. For sound waves traveling in the same direction as the source (those you hear when the vehicle approaches), the wavelength is reduced, more crests strike your ear each second, and you hear a higher frequency. As the automobile departs, you hear waves with an elongated crest-to-crest distance, giving a lower frequency.

Doppler's theory applied equally well to waves of light. It predicted that light should shift toward the red—toward longer wavelengths—when its source recedes from the observer, and toward the blue when the source approaches. In 1868 Huggins found that the spectrum of the star Sirius was shifted toward the red, by an amount that showed it to be moving away from us at 30 miles per second. His instruments could not analyze the dim nebulas, but others took up the work. By 1914 redshifts had been measured for many nebulas, indicating they were receding from us at hundreds of miles per second.

The full meaning of the nebulas and their redshifts remained elusive until Edwin Powell Hubble determined that they lay immensely far from Earth and began to explore that distant rim of creation. Hubble had graduated from the University of Chicago, where he was an excellent athlete, and attended Oxford University as a Rhodes Scholar. Then he taught school but soon abandoned it, deciding that even if he were "second-rate or third-rate, it was astronomy that mattered." He came to Mount Wilson in 1919, immediately after its 100-inch mirror had been installed as the biggest in the world. His intense dedication may explain why many saw him as aloof. Hours spent in a cold observatory, fighting the need to shiver so he would not disturb the telescope, did not encourage social chitchat. In those hours, Hubble classified the star-bearing nebulas by shape, including the spectacular spirals and elliptical forms. Most important, he established that they lay far outside our galaxy. Hubble transformed Andromeda and the multitude of other nebulas with starlike spectra into distant galaxies, each an enormous group of stars like the Milky Way.[14]

Astronomical distance such as the range to Andromeda is no trivial thing to measure. In daily life, experience and visual clues tell us how far away a light source lies. But gazing at faint images swimming in darkness, how could Hubble tell if they were small and nearby or large and distant? Did they even lie within our own galaxy? Others had reported inconsistent results, and the size of the Milky Way was not known with confidence. A scientific debate held in 1920 summarized the opposing evidence. One viewpoint held that the Milky Way was the entire universe. Perhaps 100,000 light years across, it included Andromeda 30,000 light years away, as well as the other nebulas. But some astronomers believed the Milky Way was only a tenth that large, and the spiral nebulas were distant "island universes."

Hubble resolved the debate by examining flashing stellar beacons.

Ten years earlier, Henrietta Leavitt, of the Harvard Observatory, had analyzed stars called Cepheids (named after their brightest example in the constellation Cepheus) that periodically change their luminosity in a matter of days. She found a firm relationship between the period of waxing and waning and the brightness; if you knew the cyclic rate of the star, you knew its radiant power. And just as a flashlight looks dimmer the farther you stand from it, the perceived brightness of a Cepheid would depend on its distance. The delicate linkage among period, absolute and apparent brightness, and distance depended on finding Cepheids whose distance could be established by other means, but once that fundamental calibration was made, the result was a marvelous tool. If you saw a flashing Cepheid, you were looking at a clearly marked cosmic tape measure.[15]

With the 100-inch telescope, Hubble sought and found pulsating Cepheids in the nebulas and took dozens of photographs to determine their periods. He carefully marshaled his evidence, and in 1924 he announced the astonishing conclusion: Andromeda, Messier's old M31, lay nearly a million light years from us. Hubble had carried humanity into what he called the "realm of the nebulae," moving out the boundaries of the universe. Those boundaries lay even further than he first thought: The distance to Andromeda was remeasured during World War II, when the lights of Los Angeles were blacked out as a precaution against air raids. That eliminated the glow which plagued observation at Mount Wilson and led to unusually good data that showed Andromeda to lie more than two million light years away.

Hubble's discovery made him famous. He was photographed with Hollywood stars and with Walt Disney as well as with the noted biologist Sir Julian Huxley, and he appeared on the cover of *Time* magazine. The fame was doubly deserved, because his result did more than overwhelm us with the size of the universe; it led to a second search that uncovered cosmic change and eventually cosmic birth. The idea of a changing universe came from Einstein's General Relativity of 1916, which described the spacetime of the universe and the gravity it engenders. The theory predicted that spacetime was a dynamic, changing continuum that would carry along all things in the cosmos. If so, there would be a relation between the distance to an island universe and its velocity.

Hubble explored this possibility by relating his measured distances to redshift data. Some of the redshifts were measured by Milton Humason, who had worked as a mule driver during the construction of Mount Wilson. Perhaps seeing an echo of his own commitment, Hubble recognized Humason's extraordinary self-taught enthusiasm for astronomy and set him to measuring the redshifts of galaxies up to 100 million light years away. Enthusiasm was essential, and patience as well, because these dim objects required photographic exposures up to fifty hours long. In 1929 and 1931 Hubble published the results: The velocity of recession was proportional to distance, so if one galaxy lay twice as far as another, it receded twice as fast. For the furthest galaxies, the speed was a sizable fraction of the speed of light.

That connection between speed and distance implies an expanding universe, as can be seen in a seemingly irreverent model. Picture the sweep of the cosmos as the surface of a child's balloon, sprinkled with pasted-on polka-dots. As you blow up the balloon, its radius and surface area steadily increase, as does the distance between any two dots. If you could stand on a dot, you would see every other dot moving away from you. A dot twice as far from you would recede twice as fast; the speed at which it flees is proportional to its distance. The surface of the balloon is the expanding universe, each dot is a galaxy embedded in spacetime, and the relation between recessional speed and distance is Hubble's law.

Hubble's constant—the ratio between the velocity of recession of a galaxy and its distance—became a fundamental number of cosmology. Combined with the Big Bang theory, it gave the age of the universe. But first it was essential to confirm that relativity truly predicted a changing universe. Curiously enough, Einstein himself had initially rejected the possibility. Although his theory gave a dynamic result, he put great weight on other principles that pointed to a static cosmos.

He arbitrarily added to the theory a kind of cosmic force that brought the universe back to stasis, like a balloon that neither shrinks nor grows. The Russian mathematician Aleksandr Friedmann, however, pursued the original version and in 1922 worked out the details of the dynamic universe it described. After Hubble's result, Einstein called his early preference for a fixed cosmos the "greatest blunder" of his life.[16]

Relativity gracefully accommodated an expanding universe, but it did not explain how the growth began. The idea of the Big Bang, the puff that inflated the cosmic balloon, came in 1927 from the Belgian astronomer and cleric Georges Lemaître. (The name "Big Bang" came twenty years later when the British astronomer Fred Hoyle, who had an opposing theory, first used it pejoratively.)[17] Lemaître ran in reverse a mental film of the cosmic expansion, and he watched galaxies stream back in time and space until they met in a hot "primeval atom."[18] That was the origin of the idea that the universe began with an explosion from an infinitesimal seed, which physicists call a "singularity." The receding galaxies were carried along by the universe flinging itself into nothingness after the Bang, as time, space, and all creation grew from that explosive core. And if we now see the edge of that growing volume of reality, we can tell how long the growth has taken. The mathematical reciprocal of Hubble's constant is the distance reached by the expanding rim of the universe divided by the speed of its expansion—the time it took to get there, the veritable age of the universe.

Cosmic age was one factor to weigh in judging the validity of the Big Bang theory. Early values for Hubble's constant gave an age of two billion years. That could not be right, for it would make the universe younger than the well-established age of earthly rocks, four to five billion years.[19] Later results for Hubble's constant gave more reasonable ages, up to 20 billion years. Other measurable evidence for the Bang turned out to lie in light, as predicted by the young physicists Ralph Alpher and Robert Herman. They worked with the theorist George Gamow, known for his popular books that influenced hordes of school-age scientists-to-be (including myself). In 1948 the three drew on new laboratory data for interactions among atomic nuclei to describe the behavior of nuclei soon after the Bang. In a general way, this resembled what happens inside a star, where hydrogen nuclei that repel each other are made to fuse into helium by extreme temperatures. Similarly, if elementary particles were to join into the known universe, the Bang had to be stupendously hot, but not for too long, or the new structures would decompose.[20]

From this, Alpher and Herman predicted that the dense stew of photons in the hot "primeval atom" would dilute as the universe expanded and the temperature decreased. In the present era, they wrote, the cosmos is filled with a thin soup of radiation representing an extremely low temperature, about five degrees above absolute zero. According to Planck's law of radiation, such a temperature produces invisible light over wavelengths from a few micrometers to several centimeters, or from the near infrared to the short radio waves called microwaves. The strongest emission would lie near 100 micrometers, in the far infrared region. But such radiation could not be easily detected by the technology of the time. The calculation was a surmise without a confirming experiment, soon to be nearly forgotten.

Fifteen years later, Arno Penzias and Robert Wilson stumbled across the pervasive background light, just as Herschel found Uranus. Like him, they were rewarded, with a shared Nobel Prize in 1978. In 1963 Penzias and Wilson began working with a large horn-shaped microwave antenna at Bell Laboratories in New Jersey. They were trained in radio astronomy, a science that began in 1932 when radio emissions from our own galaxy were found accidentally. Since then, radio waves have been detected from different astronomical sources. In 1951 a signal was found at a wavelength of 21 centimeters that came from atoms of hydrogen, a marker of this ubiquitous cosmic building material that has ever since provided a background "hiss" for cosmic radio reception.

Before exploring radio emissions in the sky, Penzias and Wilson had a pragmatic assignment—to minimize undesired signals in the antenna as it pointed upward. The purpose was to obtain clear reception from experimental communications satellites in orbit, forerunners of today's commercial networks in space. To their surprise, the researchers found a steady signal at a wavelength of seven centimeters (about three inches) that simply would not go away. They suspected emissions from nearby New York City or from some specific astronomical object, but the signal never varied, no matter where they pointed the antenna, no matter the time of day or season of the year.

Penzias and Wilson recognized that they were detecting the cosmic background radiation, as they announced in 1965. The uniformity of the radiation across the sky seemed to represent an initial explosion occurring everywhere at once. The measurement at a single wavelength matched a space-filling temperature of about three degrees above absolute zero, near the predicted value. Further corroboration required data over a range of far infrared and microwave wavelengths. That was difficult, because much of this light is absorbed in our atmosphere; but in the 1970s, instruments carried above most of the atmosphere in giant helium balloons confirmed that the light behaved as expected over a range of wavelengths.

That light is our most compelling evidence that the Big Bang occurred (although it is not the only evidence), and it tells us more. If the light is absolutely homogeneous, so was the early universe, but that is inconsistent with the large-scale distribution of the matter we see. Herschel was the first to note that cosmic matter is sprinkled unevenly through space. We know now that galaxies come in clusters separated by enormous voids. Superclusters of galaxies form the surfaces of gargantuan empty bubbles a hundred million light years across. And there is the "Great Wall" found in 1989: This galaxy-filled region 500 million light years long, 200 million wide, and 15 million thick, seems to be only one of many similar concentrations.[21] The early arrangement of matter that evolved into these megastructures should have produced minute variations in the background radiation across the sky. To measure the variations would require delicate observations that could only be made from space. That was the motivation for the launching in 1989 of COBE, the orbiting Cosmic Background Explorer.

COBE was not the first observatory in space. Many others have been launched since the 1960s, to ride above the blanket of air that obscures ground-based observation of the universe. Infrared and microwave photons are absorbed by atmospheric water vapor, and the atmosphere absorbs ultraviolet and X-ray light below 300 nanometers—fortunate for human survival but not for wide-range cosmic snapshots. Air pollution and the artificial glow of civilization also interfere. (This is why observatories are put high up in remote locales with dry climates.) Further, our turbulent atmosphere randomly shifts light rays as they cut through it, the reason stars twinkle. That smears out images, as a shiny coin seen through agitated water seems to move, and reduces the accuracy of observations. There is also sky glow, a faint light in the upper atmosphere due to charged particles from the sun. That makes it hard to see dim astronomical objects, as if trying to discern a candle flame in a lighted room.

The observatories floating above our atmosphere avoid these problems and can explore the universe at all wavelengths and with high resolution. The International Ultraviolet Explorer (IUE), launched in 1978, captures light from 115 to 320 nanometers with its eighteen-inch mirror; EUVE, the Extreme Ultraviolet Explorer launched in 1992, operates over wavelengths from 6 to 50 nanometers. IUE has obtained thousands of spectra, some from the intriguing objects called quasars, or quasi-stellar objects. These are the brightest sources known, emitting more light than whole galaxies yet appearing as starlike points. Their huge redshifts correspond to velocities near the speed of light, if they indeed represent velocities, and therefore enormous distances and early cosmic times. They may represent the first phases of galactic evolution. At shorter wavelengths, the Einstein Observatory of 1978 and others are showing that almost every type of astronomical object emits X rays, some at immense powers.[22]

Important as these results are, those from COBE function on a grander scale: They further confirm the Big Bang and begin to explain the look of the universe. COBE was built under the leadership of the physicist George Smoot. He earned a doctorate in elementary particle physics at MIT in 1970, but he felt a junior person could have little impact in the large teams endemic to that scientific area. Instead, he studied the cosmic background radiation. His early measurements from helium balloons and high-flying aircraft were a prelude to space-borne observation. Sensitive measurements are difficult in the microwave and far infrared regions, as I know from my own research, and the need to work in space added new constraints. COBE, which was funded by NASA, had to be redesigned after the explosion of the Challenger space shuttle required the use of a different launching vehicle. Smoot ended up after all with a large enterprise; COBE required thirteen years of effort that involved more than one thousand people. Compare this team to Edwin Hubble and his lone assistant and you see that it now takes enormous effort to explore cosmic light.[23]

COBE has performed exquisitely. Its survey of wavelengths ranging from 100 micrometers to one centimeter decisively confirms that space is filled with light corresponding to a temperature of 2.7 degrees. And data obtained after two years of observation from COBE show that primal light is not uniform. In 1992 Smoot presented a portrait of the sky, color-coded in red and blue to reflect the brightness of the background radiation. In what amounts to a map of the newborn cosmos, its colored patches show temperature differences of one part in a hundred thousand. Those hot and cold spots represent variations in the density of the early universe that, pulled and pushed by gravity, may have blossomed into the galactic clusters we see today.[24]

Although it is the long-wavelength background light that is intimately connected to the early universe, light at shorter wavelengths carries other cosmic insights, which the Hubble Space Telescope is revealing. Launched in 1990 into an orbit 380 miles above the Earth, and producing immediate frustration because its mirror had been incorrectly assembled, it is now repaired and working. The 94-inch mirror is smaller than Mount Wilson's but is enormously effective in space. Hubble's sensors operate from the ultraviolet through the visible and into the infrared. They register, in images and spectra, distant objects at high resolution. For example, Hubble sees the farthest planet Pluto and its moon Charon as crisp, well-separated disks, whereas to an earthly telescope they are diffuse disks nearly blended together.[25]

The pure light gathered by Hubble is answering fundamental questions, such as whether black holes really exist. Astronomers have long suspected that the huge elliptical galaxy M87, 50 million light years away in the constellation Virgo, harbors a black hole at its center. Now Hubble has combined pictorial light and analytical light into strong evidence for that surmise. Its telescope shows a glowing disk of gas near the center of M87, and its spectroscopic instruments show the disk to be rotating, for light from one side is blueshifted and that from the other side redshifted—one side approaches as the other recedes. The edge of the disk travels at a million miles an hour, and the disk could not remain stable without an enormous gravitational presence at its center, a mass two to three billion times that of our sun. Such a mass is far too large to come from the stars near the center of M87, making a black hole the likely candidate.[26]

The Hubble telescope is also refining the value of Hubble's constant, which in the young science of cosmology has yet to be measured with remotely the accuracy of the speed of light. Some ground-based measurements give a value corresponding to a cosmic age of 20 billion years, but other methods and interpretations yield values up to twice as large, which translate into ages of 10 to 12 billion years. Images from Hubble show Cepheid variables in the extremely distant group of galaxies in the constellation Virgo, which are thought to give especially meaningful results. The first data, reported in 1994, give an age of 12 billion years. That is a troublesome outcome, for it is less than the age of some stars.[27]

Such difficulties only indicate how much remains to be learned from universal light. Even if we knew Hubble's constant exactly, we must also know the amount of cosmic matter in order to find a meaningful age for the universe and to predict its final behavior. As galaxies race out into space and time, their gravitational interaction pulls them back, slowing the expansion. The more material in the universe, the greater the tug. At a certain critical density, equivalent to a few atoms of hydrogen in a room-size volume, the cosmos is balanced just between expansion and contraction. In the theory, the ratio of the actual density of matter to that critical value is represented by Ω (omega), the last letter of the Greek alphabet—an appropriate symbol, for it carries the fate of the universe. If Ω is less than or equal to one, the cosmos expands forever, more or less rapidly; if Ω exceeds one, there is enough mass to eventually make the cosmos contract, the process opposite to the Big Bang called the Big Crunch.[28]

It is remarkable that a single number contains so much meaning and that Ω can be determined in principle by carefully counting up all the galaxies, dust, and gas we see in space. But there is a complication. As much as 90 percent of cosmic matter is indifferent to light, neither absorbing nor emitting it. We infer the existence of this "dark matter" from the gravitational and rotational behavior of galaxies, but its constitution is a mystery.

It may consist of exotic new elementary particles, yet to be found; or of the known but elusive elementary particles called neutrinos; or of burned-out stars, or invisible black holes. Like any matter, however, dark matter affects light through the gravity it produces. The deflection of the image of a star by our sun's gravity, seen in 1919, was the first confirmation of general relativity. Similarly, the gravity from a cluster of dark matter changes the appearance of a distant source as its light passes. Such "gravitational lenses" are now being used to map the amounts and shapes of dark matter, so that light examines even this invisible component of the universe.[29]

At present, there are reasons to believe that when all the matter is counted up, Ω is near one.[30] But new measurements are only beginning to establish an accurate quantitative cosmology. Nothing is ruled out as yet, including the evocative possibility of the Big Crunch. At first glance, that represents final closure, but some believe it represents a beginning as well, for as the cosmos crashes inward, it may contract into a new primordial atom that will grow into a new version of reality—a quasi-cyclic shift from Bang to Crunch to Bang.

Whether the Crunch is cosmic finale or part of a cycle, it adds meaning to universal light, which already has dual meaning. It is the light that defines the universe, its flow from the far edge of creation our access to distant times and environs. All we know of the Big Bang, the Great Wall, the age of the cosmos, traces back to a scattering of photons that alters a handful of molecules in our eyes, darkens a few grains of silver in a photographic emulsion, releases a trickle of electrons in a silicon detector. Universal light is also ubiquitous light, all-reaching light. We look outward to the universe and find it filled with the photonic spoor of the Big Bang. We gaze inward, and there again is light, its perception occupying the brain and informing the mind. It is what enables clusters of neurons to weigh and measure galaxies of stars.

The possibility of a Big Crunch gives a third meaning to universal light. If there were a time when the universe reached its limit of expansion, when the outflowing galaxies stopped, quivered, and reversed themselves, only light would signal the beginning of that new era. Were there any living beings to observe, at that instant cosmic color would for once become absolutely true, for the redshift would cease, and then the colors of galaxies would shift toward the blue, as each began its inward race.

That image inspires me to carry the fantasy further. During a Big Crunch, there might be local variations as the universe recondensed, as if the deflating cosmic balloon remained taut here, became limp and wrinkled there. I envision light waves as shrinking in some parts of the universe, perhaps in our own Milky Way. That is valid only in my imagination, not in any physical theory I know, but the human meaning of light becomes clear when I contemplate the havoc such a change would play. As that terrible shrinkage left behind wavelength after wavelength, it would strip away layer after layer of sense contact with the universe, of understanding and communication. First to be lost would be the long wavelengths of radio, microwaves, the far infrared. Gone would be the hiss of radio stars, the twenty-one-centimeter signal from hydrogen, word and image carried by radio and television transmissions. As the middle and near infrared were left behind, the warmth of sunlight on skin would cease, the great fiber optic network would fail, reproduced music would fall silent.

Next the fringes of vision would blur and vanish, beginning with the color red. Here would dim the last glow of cadmium red and orange and yellow, of Van Gogh's beloved sunflowers. Sea green would depart, followed by the blues—gone the varied shades of light scattered in the sky, the delicate blue-violet of Monet's *Water Lilies*. Only the plainest art would survive. Etchings or photographs in black and white, silent films and film noir, old television kinescopes—so long as there were some visible light their outlines would carry meaning even in a monochrome world. The ultraviolet wavelengths would fade next, taking with them fears of sunlight as the agent of cancer. Soon after, the harm due to X rays would be gone, but so would their power to diagnose human ills. As shorter wavelengths yet were left behind, fear of nuclear radiation would diminish, for its destructive gamma rays lie in this region. But what could those few benefits mean in a universe devoid of light and warmth? Could so bleak a cosmos even hold gods or a God?

The Crunch is not here yet, if indeed it will ever come. At worst, a dark and barren universe lies unimaginably far in the future. We live still in a world flooded with photons and waves, with light and color, with an ethereal energy we do not fully understand, that defines what we are and what we know—an empire of light that holds sway in space, time, and meaning.

Notes

1. The Birth and Meaning of Light

Epigraph: Hazel Rossotti, *Colour: Why the World Isn't Grey* (Princeton, N.J.: Princeton University Press, 1983), 229.

2. Seeing Light

Epigraph 1: Cited in John Bartlett, *Bartlett's Familiar Quotations* (Boston: Little, Brown and Company, 1980), 1,040.

Epigraph 2: Santiago Ramón y Cajal, *Recollections of My Life,* E. Horne Craigie, translator. (Philadelphia: The American Philosophical Society, 1937), 576.

1.   George Sperling, "Comparison of Perception in the Moving and Stationary Eye," in Eileen Kowler, ed., *Eye Movements and Their Role in Visual and Cognitive Processes* (Amsterdam: Elsevier, 1990), 311. My discussion of the different aspects of visual perception draws on Kowler; Hazel Rossotti, *Colour: Why the World Isn't Grey* (Princeton, N.J.: Princeton University Press, 1983), 109–166; Irvin Rock, *Perception* (New York: Scientific American Library, 1984), 1–13, 100–105, 145–151; John E. Dowling, *The Retina: An Approachable Part of the Brain* (Cambridge, Mass.: Belknap Press, 1987); Julie I. Schnapf and Denis A. Baylor, "How Photoreceptor Cells Respond to Light," *Scientific American,* April 1987, 40–47; Lubert Stryer, "The Molecules of Visual Excitation," *Scientific American,* July 1987, 42–51; David H. Hubel, *Eye, Brain, and Vision* (New York: Scientific American Library, 1988), 1–11, 59–125; Nicholas J. Wade and Michael Swanston, *Visual Perception: An Introduction* (London: Routledge, 1991), 16–39, 59–95; Daniel C. Dennett, *Consciousness Explained* (Boston: Little, Brown, 1991), 52–60, 111–138, 383–389; Rajinder P. Khosla, "From Photons to Bits," *Physics Today,* December 1992, 42–49.

2.   Guy Thomas Buswell, *How People Look at Pictures: A Study of the Psychology of Perception in Art* (Chicago: University of Chicago Press, 1935), 1–92, 142–194.

3.   Lael Wertenbaker, *The Eye: Window to the World* (New York: Torstar Books, 1984), 41.

4. Beth Ann Meehan, "Seeing Red: It's Written in Your Genes," *Discover,* June 1993, 66.

5. Hubel, 8.

6. Semir Zeki, "The Visual Image in Mind and Brain," *Scientific American,* September 1992, 69–76; Denis Grady, "The Vision Thing: Mainly in the Brain," *Discover,* June 1993, 62–64.

7. Rock, 235.

8. Francis Crick and Christof Koch, "The Problem of Consciousness," *Scientific American,* September 1992, 153–159.

9. Diane Ackerman, *A Natural History of the Senses* (New York: Vintage, 1990), 254.

10. Jan B. Deregowski, "Pictorial Perception and Culture," *Scientific American,* November 1972, 82–88; Rock, 103–105.

11. Oliver Sacks, "To See or Not to See," *The New Yorker,* 10 May 1993, 59–73.

12. Rock, 145.

13. Richard Kendall, "Degas and the Contingency of Vision," *The Burlington Magazine,* March 1988, 180–197.

14. Kendall, 190.

15. Grady, 62–64.

16. Alfred Bester, *The Stars My Destination* (Boston: Gregg Press, 1975), 130.

3. Classical Light

Epigraph: Cited in Gale E. Christianson, *In the Presence of the Creator: Isaac Newton and His Times* (New York: The Free Press, 1984), 44; which notes that the lines were carved above the fireplace in the room where Newton was born. According to Derek Gjersten, *The Newton Handbook* (London: Routledge and Kegan Paul, 1986), 439; Pope wrote an earlier version of the lines in 1727 for a monument to Newton to be placed in Westminster Abbey, but the epitaph (in Latin) "Who surpassed all men in genius" was used instead.

1. My discussion of theories of light from the classical world through the Middle Ages draws on Vasco Ronchi, *The Nature of Light,* V. Barocas, translator (Cambridge, Mass.: Harvard University Press, 1970), 1–76; Eugene Hecht and Alfred Zajac, *Optics* (Reading, Mass.: Addison-Wesley, 1979), 1–10; Arthur Zajonc, *Catching the Light* (New York: Bantam Books, 1993), 19–22, 77–78, 262–264; Florian Cajori, *A History of Physics* (New York: Dover Publications, 1962), 9–12, 28–29.

2. Ronchi, 15.

3. Cajori, 23.

4. Christian Huygens, "The Wave Theory of Light," in Holmes Boynton,

editor, *The Beginnings of Modern Science: Scientific Writings of the 16th, 17th, and 18th Centuries* (Roslyn, N. Y.: Walter J. Black, 1948), 137–146.

5. My discussion of Newton's life and work is based on Richard S. Westfall, *The Life of Isaac Newton* (Cambridge: Cambridge University Press, 1993).

6. H. W. Turnbull, editor, *The Correspondence of Isaac Newton, Volume 1, 1661–1675* (Cambridge: Cambridge University Press, 1959), 169.

7. Westfall, 81–84.

8. Gjersten, 507–508.

9. Isaac Newton, "A New Theory of Light and Colors," in Boynton, 148–156.

10. Isaac Newton, *Opticks* (New York: Dover Publications, 1952), 400.

11. Anthony Blunt, "Blake's 'Ancient of Days': The Symbolism of the Compasses," in Robert N. Essick, editor, *The Visionary Hand: Essays for the Study of William Blake's Art and Aesthetics* (Los Angeles: Hennessey and Ingalls, 1973), 71–125; Robert N. Essick, *William Blake, Printmaker* (Princeton, N.J.: Princeton University Press, 1980), 37–38.

12. Works of art portraying Newton are listed in William Cooledge Lane and Nina E. Browne, editors, *American Library Association Portrait Index* (Washington, D.C.: Government Printing Office, 1906), 1,270. Several works, including Pittoni's *Allegorical Tomb for Isaac Newton,* are shown and discussed in Francis Haskell, "The Apotheosis of Newton in Art," Robert Palter, editor, *The Annus Mirabilis of Sir Isaac Newton, 1666–1966* (Cambridge, Mass.: M.I.T. Press, 1970), 302–321. A larger image of *Allegorical Tomb,* along with details of the painting and its other versions, appears in Franca Zava Boccazi, *Pittoni: L'Opera Completa* (Venice: Alfieri, 1979), figures 202–208. The comment about Romney's portrait appears in Haskell, 318; and the portrait *Newton and the Prism* can be seen in Martin Kemp, *The Science of Art* (New Haven: Yale University Press, 1990), 293.

13. Haskell, 318. Views of the proposed cenotaph appear in Fritz Wagner, *Isaac Newton im Zwielicht zwischen Mythos und Forschung* (Freiburg: Alber, 1976), plates 8–10. It is compared to the pyramid at Giza, the domed Nazi hall, and other structures in Albert Speer, *Albert Speer: Architecture, 1932–1942* (Brussels: Archives d'Architecture Moderne, 1985), 12.

14. William Wordsworth and Ernest de Selincourt, editor, *The Prelude* (Oxford: Clarendon Press, 1959), 75.

15. Ryan D. Tweney, "Fields of Enterprise: On Michael Faraday's Thought," in Doris B. Wallace and Howard E. Gruber, *Creative People at Work* (New York: Oxford University Press, 1989), 90–126.

16. My discussion of Michelson's life and work is based on Bernard Jaffe,

Michelson and the Speed of Light (Garden City, N.Y.: Anchor Books, 1960) and Dorothy Michelson Livingston, *The Master of Light: A Biography of Albert A. Michelson* (New York: Charles Scribner's Sons, 1973). The reference to his sea duty and its relation to his thinking about motion appears in Livingston, 38–40.
17. Jaffe, 88–90.

4. Modern Light

Epigraph: Cited in Emil Wolf, "Einstein's Researches on the Nature of Light," *Optics News,* vol. 5 (1979), 24.

1. Linda Dalrymple Henderson, *The Fourth Dimension and Non-Euclidean Geometry in Modern Art* (Princeton, N.J.: Princeton University Press, 1983), 58.

2. Henderson, 356, quoting Einstein in a letter to Paul M. Laporte, who had proposed a definite relationship between cubism and Special Relativity. For a complete recounting of how the two mistakenly came to be intertwined, see Henderson, 353–365. See also Michael Baxandall, *Patterns of Intention: On the Historical Explanation of Pictures* (New Haven, Conn.: Yale University Press, 1985), 76.

3. My discussion of Einstein's life and the origins of the theory of relativity draws on Paul Arthur Schilpp, editor, *Albert Einstein: Philosopher-Scientist* (Evanston, Ill.: The Library of Living Philosophers, 1949); Albert Einstein, *Relativity: The Special and the General Theory* (New York: Wings Books, 1961); Albert Einstein, *Autobiographical Notes,* P. A. Schilpp, translator and editor (La Salle, Ill.: Open Court Publishing, 1979); Arthur I. Miller, "Imagery and Intuition in Creative Scientific Thinking: Albert Einstein's Invention of the Special Theory of Relativity," in Doris B. Wallace and Howard E. Gruber, *Creative People at Work* (New York: Oxford University Press, 1989), 171–187. The quotes "indescribable impression" and "pure thinking" come from Einstein (1979), 9, 11.

4. Einstein (1979), 7.

5. See, for instance, Bernard Jaffe, *Michelson and the Speed of Light* (Garden City, N.Y.: Anchor Books, 1960), 100–101, 167–168; Paul A. Tipler, *Physics for Scientists and Engineers* (New York: Worth, 1991), 1,106.

6. The photograph appears in John Szarkowski, *The Photographer's Eye* (New York: Museum of Modern Art/Doubleday, 1966), 85.

7. J. C. Hafele and Richard E. Keaton, "Around the World Atomic Clocks: Predicted Relativistic Time Gains," *Science,* vol. 177 (1972), 166–168; "Around the World Atomic Clocks: Observed Relativistic Time Gains," *Science,* vol. 177 (1972), 168–170.

8. David Deutsch and Michael Lockwood, "The Quantum Physics of Time Travel," *Scientific American,* March 1994, 68–74; Ivars Peterson, "Chaos in Spacetime," *Science News,* vol. 144 (1993), 376–377.

9. Bruce R. Wheaton, *The Tiger and the Shark: Empirical Roots of Wave-Particle Dualism* (Cambridge: Cambridge University Press, 1983), 106. Much of my discussion of the early days of the photon and the wave-particle duality follows Wheaton; also Dipankar Home and John Gribbin, "What Is Light?" *New Scientist,* 2 November 1991, 30–33; and J. P. Vigier, "From Descartes and Newton to Einstein and de Broglie," *Foundations of Physics,* vol. 23 (1993), 1–4.

10. Wheaton, 278, 280.

11. Cited in Wheaton, 109.

12. Rodney Loudon, *The Quantum Theory of Light* (Oxford: Clarendon Press, 1981), 229–230; Arthur Zajonc, *Catching the Light* (New York: Bantam Books, 1993), 297–298.

13. G. I. Taylor, "Interference Fringes with Feeble Light," *Proceedings of the Cambridge Philosophical Society,* vol. 15 (1909), 114–115. Peter Mason, in *The Light Fantastic* (Ringwood, Victoria, Australia: Penguin Australia, 1981), 81–82, estimates that Taylor actually measured the impact of only one photon at a time.

14. P. Grangier, G. Roger, and A. Aspect, "Experimental Evidence for a Photon Anticorrelation Effect on a Beam Splitter: A New Light on Single-Photon Interferences," *Europhysics Letters,* vol. 1 (1986), 173–179.

15. Wheaton, 291, 299.

16. Richard P. Feynman, *QED: The Strange Theory of Light and Matter* (Princeton, N.J.: Princeton University Press, 1985), 37; Richard P. Feynman, Robert B. Leighton, and Matthew Sands, *The Feynman Lectures on Physics, Vols. I, II and III* (Reading, Mass.: Addison-Wesley, 1966), vol. 3, 1.1.

17. Cited in John Bartlett, *Bartlett's Familiar Quotations* (Boston: Little, Brown and Company, 1992), 636.

18. Einstein (1979), 49, 83.

19. Cited in Hans Christian von Baeyer, *Taming the Atom* (New York: Random House, 1992), 221.

20. Cited in Wheaton, 306.

21. David H. Freedman, "Theorists to the Quantum Mechanical Wave: 'Get Real,' " *Science,* 12 March 1993, 1542–1543.

22. Delayed-choice experiments and their interpretation are discussed by John Horgan, "Quantum Philosophy," *Scientific American,* July 1992, 94–101. For a scientific report on one such experiment, see T. Hellmuth, H. Walther, A. Zajonc, and W. Schleich, "Delayed-Choice Experiments in Quantum Interference," *Physical Review A,* vol. 35 (1987), 2,532–2,541.

23. Cited in article "William of Ockham," William H. Harris and Judith S.

Levy, editors, *The New Columbia Encyclopedia* (New York: Columbia University Press, 1975), 2,981.

24. Georges Braque, *Cahier de Georges Braque, 1917–1947* (Paris: Maeght, 1948), 10.

25. Braque, 52.

5. Making Light

Epigraph: Quoted in Christopher Cerf and Victor Navasky, *The Experts Speak* (New York: Pantheon, 1984), 203.

1. These descriptions come from James W. Shepp and Daniel B. Shepp, *Shepp's World's Fair Photographed* (Chicago: Globe Bible Publishing, 1893); Carolyn Marvin, *When Old Technologies Were New* (New York: Oxford University Press, 1988), 171–173. The quotes are given in Shepp and Shepp, 8, 58.

2. The anecdote appears in L. Sprague de Camp, *Heroes of American Invention* (New York: Barnes and Noble, 1993), 168–169.

3. My discussion of the early days of illumination draws on William T. O'Dea, *The Social History of Lighting* (London: Routledge and Kegan Paul, 1958), 1–65, 153–176, 213–232; Leroy Thwing, *Flickering Flames* (Rutland, Vt.: Charles E. Tuttle, 1958), 1–28, 71–99; Peter Mason, *The Light Fantastic* (Ringwood, Victoria, Australia: Penguin Australia, 1981), 9–26; Wolfgang Schivelbusch, *Disenchanted Night: The Industrialization of Light in the Nineteenth Century* (Berkeley and Los Angeles: University of California Press, 1988). The quote comes from Schivelbusch, 81.

4. Mario Ruspoli, *The Cave of Lascaux* (New York: Harry N. Abrams, 1987), 28–30.

5. L. Sprague de Camp, *The Ancient Engineers* (New York: Ballantine, 1962), 258; Mason, 10–11; O'Dea, 31.

6. The earliest definite evidence for the candle dates to 1 A.D., according to O'Dea, 17. The discussion of origins in Africa and among the Phoenicians comes from the article "Lighting and Lighting Devices," *The New Encyclopædia Britannica,* 15th edition (Chicago: Encyclopædia Britannica, 1991), vol. 23, 29–38; the attribution to the Etruscans is made in De Camp (1962), 174. Apparently difficulties in translation make it unclear whether early writings refer to "lamps," "torches," or "candles," which complicates assignments of time and place of origin.

7. David Brenner and Ann Prescott, "Painting with Light," *New Scientist,* vol. 102 (1984), 40–41.

8. Michael Faraday, "The Chemical History of a Candle," in Mary Elizabeth Bowen and Joseph A. Mazzeo, *Writing About Science* (New York: Oxford University Press, 1979), 7–19.

9. Simon Schama, *Citizens: A Chronicle of the French Revolution* (New York: Alfred A. Knopf, 1989), 72–79, 313.

10. Thwing, 72.

11. Thwing, 58.

12. Schivelbusch, 151; Schama, 135.

13. De Camp (1962), 352.

14. W. J. Passingham, *Romance of London's Underground* (New York: Benjamin Blom, 1972), 19–20; Benson Bobrick, *Labyrinths of Iron* (New York: Newsweek Books, 1981), 102.

15. Schivelbusch, 55, 115.

16. Schivelbusch, 3, 128–134.

17. O'Dea, 100–102.

18. My discussion of Edison's life and work draws on De Camp (1993), 168–193; Robert Friedel and Paul Israel, *Edison's Electric Light* (New Brunswick, N.J.: Rutgers University Press, 1986).

19. Friedel and Israel, 6, 36, 56, 116–117.

20. Article "William Crookes" in Charles Coulston Gillispie, editor, *Dictionary of Scientific Biography* (New York: Charles Scribner's Sons, 1970), vol. 3, 474–482.

21. O'Dea, 168.

22. Edward Hopper, *Edward Hopper: Forty Masterworks* (New York: W. W. Norton, 1988). Cited in introductory essay by Heinz Liesbrock, 2.

23. Hopper, plate 26.

24. Robert Hobbs, *Edward Hopper* (New York: Harry N. Abrams, 1987), 123–124, 129.

25. Hopper, plate 29.

26. Hopper, plate 22.

27. Sandor Kuthy and Kuniko Satonubu, *Sonia and Robert Delaunay* (Stuttgart: Verlag Gerd Hatje, 1991), 67, 92.

28. Anne d'Harnoncourt and Walter Hopps, *Etant Donnés: 1° La Chute d'eau; 2° Le gaz d'éclairage: Reflections on a New Work by Marcel Duchamp* (Philadelphia: Philadelphia Museum of Art, 1973), 35, 37, 39.

29. O'Dea, plate 13; Nigel Gosling, *Gustave Doré* (New York: Praeger, 1973), 91; Annie Renonciat, *La Vie et l'Oeuvre de Gustave Doré* (Paris: ACR Edition, 1983), 195.

30. O'Dea, 107; plate 17.

31. Michael Kitson, *The Complete Paintings of Caravaggio* (New York: Harry N. Abrams, 1967), 5, 6.

32. Jacques Thuillier, *Georges de La Tour* (Paris: Flammarion, 1992), 162–167, 225, 239.

33. Robert Irwin and Russell Ferguson, *Robert Irwin* (Los Angeles: Museum of Contemporary Art, 1993), 122–123.

34. James Turrell, Barbara Haskell, and Melinda Wortz, *Light and Space* (New York: Whitney Museum of American Art, 1980), 15, 20–23. The quote about "revelation" comes from James Turrell, Julia Brown, and Craig E. Adcock, *Occluded Front, James Turrell* (Los Angeles: Fellows of Contemporary Art, 1985), 43.

35. Schivelbusch, 7.

36. Schivelbusch, 84–86, 98–99, 102–104, 121–123.

37. O'Dea, 225.

38. Albert Speer, *Inside the Third Reich* (New York: Macmillan, 1970), 58–59, 528.

39. Joan Lisa Bromberg, *The Laser in America, 1950–1970* (Cambridge, Mass.: MIT Press, 1991), 69–73.

6. Shaping Light

Epigraph: G. F. Imbusch and W. M. Yen, "Ruby—Solid-State Spectroscopy's Serendipitous Servant" in W. M. Yen and M. D. Levenson, editors, *Lasers, Spectroscopy, and New Ideas* (Berlin: Springer-Verlag, 1987), 248.

1. E. Liddall Armitage, *Stained Glass: History, Technology, and Practice* (Newton, Mass.: Charles T. Branford, 1959), 20–21.

2. Robert L. Feller, *Artists' Pigments: A Handbook of Their History and Characteristics* (Washington, D.C.: National Gallery of Art, 1986), 255, 256, 280.

3. This translation is cited in Wilfred Niels Arnold, *Vincent van Gogh: Chemicals, Crises, and Creativity* (Boston: Birkhauser, 1992), 221. It comes from a letter Van Gogh wrote to his sister Wilhelmina, which appears as letter W7 in Maurice Beerblock, Louis Roëlandt, and Georges Charensol, *Vincent van Gogh: Correspondance Générale* (Paris: Gallimard, 1990), 284.

4. A detailed discussion of the pigments is given in Vojtech Jirat-Wasiutynski and H. Travers Newton, Eugene Farrell, and Richard Newman, *Vincent van Gogh's "Self-portrait Dedicated to Paul Gauguin": An Historical and Technical Study* (Cambridge, Mass.: Center for Conservation and Technical Studies, Harvard University Art Museums, 1984), 28, 32–33, 36. The painting is shown there, and in Ingo F. Walther and Rainer Metzger, *Vincent van Gogh: The Complete Paintings* (Cologne: Benedikt Tascher, 1990), 396.

5. Cited from Van Gogh's letters in Irving Stone, editor, *Dear Theo* (Boston: Houghton Mifflin Company, 1937), 456.

6. Feller, 102, 104.

7. Feller, 65–102. The quote about "beauty and . . . fixity" appears on page 67.

8. Arnold, 186–192, 307. See also Natalie Angier, "New Explanation Given

for van Gogh's Agonies," *New York Times,* 21 December 1991, L11.

9. *Wheatfield with Cypresses* hangs in the National Gallery, London. It can be seen in Walther and Metzger, 544. *The Langlois Bridge at Arles* hangs in the Walraf-Richartz Museum, Cologne, and can be seen in Walther and Metzger, 343. Van Gogh painted several other images of this bridge in 1888, with the same or similar titles; they appear in Walther and Metzger, 322–325.

10. Ovid and Rolfe Humphries, translator, *Metamorphoses* (Bloomington: Indiana University Press, 1955), 70.

11. Eugene Hecht and Alfred Zajac, *Optics* (Reading, Mass.: Addison-Wesley, 1979), 2; article "Mirror," *The New Encyclopædia Britannica,* 15th edition (Chicago: Encyclopædia Britannica, 1991), vol. 8, 182.

12. L. Sprague de Camp, *The Ancient Engineers* (New York: Ballantine, 1962), 274.

13. Fred Leeman, Joost Elffers, and Michael Schuyt, *Hidden Images: Games of Perception, Anamorphic Art, Illusion from the Renaissance to the Present* (New York: Harry N. Abrams, 1976), 111–134, plates 128–167. See also Jurgis Baltrušaitis, *Anamorphic Art* (New York: Harry N. Abrams, 1977).

14. Article "John William Strutt, Third Baron Rayleigh" in Charles Coulston Gillispie, editor, *Dictionary of Scientific Biography* (New York: Charles Scribner's Sons, 1970), vol. 13, 100–108.

15. Arthur Zajonc, *Catching the Light* (New York: Bantam Books, 1993), 282–284.

16. Article "Willebrord Snel," in Gillispie, vol. 12, 499–502.

17. Florian Cajori, *A History of Physics* (New York: Dover Publications, 1962), 9.

18. De Camp, 274; Hecht and Zajac, 1.

19. Article "Roger Bacon" in Gillispie, vol. 1, 377–385; De Camp, 347–349. Bacon's comment about "sun, moon, and stars" is cited in De Camp, 349.

20. Cajori, 10.

21. *Starry Night over the Rhone* (1888) is in Paris, Musée d'Orsay; *Starry Night* (1889) resides in New York, The Museum of Modern Art; *Road with Cypress and Star* (1890) hangs in the Rijksmuseum Kröller-Müller, Otterlo. An image of each appears in Walther and Metzger, 431, 520–521, and 632, respectively. The accurate astronomy exhibited in these works is discussed by Roger Sinnott, "Astronomical Computing: Van Gogh, Two Planets, and the Moon," *Sky and Telescope,* October 1988, 406–408; Charles A. Whitney, "The Skies of Vincent van Gogh," *Art History,* September 1986, 351–362.

7. Using Light

Epigraph: Quoted in John Carey and Neil Gross, "The Light Fantastic," *Business Week,* 10 May 1993, 44.

1. My discussion of photonic technology for communication over optical fiber draws on: National Research Council, *Photonics: Maintaining Competitiveness in the Information Era* (Washington, D.C.: National Academy Press, 1988), 9–22; Terry Edwards, *Fiber-Optic Systems: Network Applications* (Chichester, West Sussex, Eng.: John Wiley and Sons, 1989), 1–11; Bahaa E. A. Saleh and Malvin Carl Teich, *Fundamentals of Photonics* (New York: John Wiley and Sons, 1991), 592–692, 874–915; Fiberoptics Components Handbook, *Laser Focus World,* June 1994, S5–S40.

2. Michael Shimazu, "Optical Computing Coming of Age," *Photonics Spectra,* November 1992, 66–74; M. Mitchell Waldrop, "Computing at the Speed of Light," *Science,* 22 January 1993, 456; W. Wayt Gibbs, "Light Motif," *Scientific American,* April 1993, 116–117; Carey and Gross, 44–50.

3. James Glanz, "DOE Lifts Veil of Secrecy from Laser Fusion," *Science,* 17 December 1993, 1,811–1,812.

4. "Nova Nears Completion," *Physics Today,* September 1984, 20; R. Stephen Craxton, Robert L. McCrory, and John M. Soures, "Progress in Laser Fusion," *Scientific American,* August 1986, 68–79.

5. Gary Taubes, "Laser Fusion Catches Fire," *Science,* 3 December 1993, 1,504–1,506; William J. Broad, "Vast Laser Would Advance Fusion and Retain Bomb Experts," *New York Times,* 21 June 1994, B7, B10; Charles T. Troy, "Laser Facility Will Seek to Duplicate the Big Bang," *Photonics Spectra,* May 1994, 31.

6. G. P. Williams, R. Budhani, C. J. Hirschmugl, G. L. Carr, S. Perkowitz, B. Lou, and T. R. Yang, "Infrared Synchrotron Radiation Transmission Spectroscopy of YBaCuO in the Gap and Supercurrent Region," *Physical Review B,* vol. 41 (1990), 4,752–4,755.

7. E. Pennisi, "Synchrotron Beam Sees Record Tiny Crystal," *Science News,* 14 September 1991, 164.

8. My discussion of research at NSLS draws on these reports published by the Brookhaven National Laboratory, and available from the National Technical Information Service: Gwyn Williams, *Using the Light Fantastic* (1985); *Brookhaven Highlights* for 1990, 1991, and 1992; S. L. Hulbert and N. M. Lazarz, editors, *National Synchrotron Light Source Annual Report 1991,* vols. 1, 2 (1992).

9. George S. Brown, "Imaging the Heart Using Synchrotron Radiation," *Beamline,* vol. 23, no. 3 (1993), 22–28.

10. Herman Winick and Gwyn Williams, "Overview of Synchrotron Radiation Sources World-Wide," *Synchrotron Radiation News,* vols. 4, no. 5 (1991), 23–26.

11. David A. Benaron and David K. Stevenson, "Optical Time-of-Flight and

Absorbance Imagining of Biologic Media," *Science,* 5 March 1993, 1,463–1,466; Joseph Alper, "Transillumination: Looking Right Through You," *Science,* 30 July 1993, 560; Kristin Leutwyler, "Optical Tomography," *Scientific American,* January 1994, 147–148.

12.   Michael M. Haglund, George A. Ojemann, and Daryl W. Hochman, "Optical Imaging of Epileptoform and Functional Activity in Human Cerebral Cortex," *Nature,* vol. 358 (1992), 668–671; Warren E. Leary, "Optical Imaging Offers Gentler Way to Monitor Human Brain at Work," *New York Times,* 25 August 1992, B6.

13.   Quoted in Jeff Hecht, "Military Lasers: The Incredible Becomes Credible," *Laser Focus World,* July 1994, 57.

14.   Joan Lisa Bromberg, *The Laser in America, 1950–1970* (Cambridge, Mass.: MIT Press, 1991), 234–238; Vincent Kiernan, "Military Lasers Face Cloudy Future," *Laser Focus World,* May 1994, 71–72; W. Conrad Holton, "The Light Brigade," *Photonics Spectra,* June 1994, 77; Jeff Hecht, "Military Lasers: The Incredible Becomes Credible." *Laser Focus World,* July 1994, 57–58, 60; William J. Broad, "From Fantasy to Fact: Space-Based Laser Nearly Ready to Fly," *New York Times,* 6 December 1994, B5–B6.

15.   "Thermal Imaging Joins Lyme-Disease Battle," *Photonics Spectra,* July 1994, 20, 22.

16.   Peter Applebome, "Sodden Midwest Is Bracing for More Rain and Floods," *New York Times,* 22 July 1993, A8.

17.   Phil Patton, "The Pixels and Perils of Getting Art on Line," *New York Times,* 7 August 1994, Arts and Leisure 1, 31.

18.   John Travis, "Laser Replication of Rare Art," *Science,* 15 May 1992, 969.

19.   Harriet Paterson, "Raphael, the Duke and £20 Million," *Sunday Telegraph,* 9 February 1992, 115.

20.   Michael Kimmelman, "A Face in the Gallery of Picasso's Muses Is Given a New Name," *New York Times,* 21 April 1994, B1, B2; William Rubin, "The Pipes of Pan: Picasso's Aborted Love Song to Sara Murphy," *ARTNews,* May 1994, 138–147.

8.   Universal Light

Epigraph: Edwin Hubble, "The Science of the Sky," in Warren Weaver, editor, *The Scientists Speak* (New York: Boni and Gaer, 1947), 40.

1.   The anecdote and the quote come from Willy Ley, *Watchers of the Skies* (New York: Viking Press, 1963), 84.

2.   My discussion of Galileo's life and work draws on J. J. Fahie, *Galileo: His Life and Work* (London: John Murray, 1903), 378–379, 381–382; Henry C. King, *The History of the Telescope* (London: Charles Griffin, 1955), 34–47;

Stillman Drake, *Discoveries and Opinions of Galileo* (Garden City, N.Y.: Doubleday, 1957); Article "Galileo Galilei," in Charles Coulston Gillispie, editor, *Dictionary of Scientific Biography* (New York: Charles Scribner's Sons, 1970), vol. 5, 237–249. The quote about wine is cited in Drake, 5.

3.　V. Greco, G. Molesini, and F. Quercioli, "The Telescopes of Galileo," in Philip P. Schewe and Ben P. Stein, editors, *American Physical Society News: Physics News in 1993,* May 1994, S16.

4.　Marian Card Donnelly, *A Short History of Observatories* (Eugene: University of Oregon, 1973), 14, 20.

5.　Richard Learner, *Astronomy Through the Telescope* (New York: Van Nostrand Reinhold, 1981), 114–115.

6.　My discussion of Herschel's life and work draws on Learner, 57–62, where Herschel's forty-foot telescope is shown on 60–61; and King, 120–142.

7.　Lord Rosse's life and work are discussed in Learner, 74–78.

8.　Charles A. Whitney, "The Skies of Vincent van Gogh," *Art History,* September 1986, 351–362.

9.　The description of Alvan Clark's grinding and polishing methods comes from Deborah Jean Warner, *Alvan Clark and Sons: Artists in Optics* (Washington, D.C.: Smithsonian Institution, 1968), 25–27. The quote about martyrdom appears on page 25.

10.　Warner, 112.

11.　Bradford L. Smith, "World's Largest Mirror to Paris," *Photonics Spectra,* September 1993, 30–31; John Noble Wilford, "New Galaxy May Shed Light on Universe," *New York Times,* 1 February 1996, A12.

12.　E. Scott Barr, "The Infrared Pioneers," *Infrared Physics,* vol. 1 (1961), 1; R. A. Smith, F. E. Jones, and R. P. Chasmar, *The Detection and Measurement of Infrared Radiation* (Oxford: The Clarendon Press, 1968), 1–2.

13.　A photograph of the spectrum can be seen in Learner, 93.

14.　My discussion of Hubble's life and work draws on Richard Berendzen, Richard Hart, and Daniel Seeley, *Man Discovers the Galaxies* (New York: Science History Publications, 1976), 199–210; Alexander S. Sharov and Igor D. Navikov, *Edwin Hubble: The Discoverer of the Big Bang Universe,* Vitaly Kisin, translator (Cambridge: Cambridge University Press, 1993). The quote "it was astronomy that mattered" appears in Berendzen, Hart, and Seeley, 206.

15.　Berendzen, Hart, and Seeley, 22–23.

16.　Cited in Marcia Bartusiak, *Through a Universe Darkly* (New York: HarperCollins, 1993), 125.

17.　Ralph Alpher and Robert Herman, "Reflections on Early Work on 'Big Bang' Cosmology," *Physics Today,* August 1988, 24–34. The anecdote appears on page 24.

18. LeMaître's visualization of the early universe is described in Bartusiak, 125. General discussions of the origin of the Big Bang theory appear in George Smoot and Keay Davidson, *Wrinkles in Time* (New York: William Morrow, 1993), 42–86; Barry Parker, *The Vindication of the Big Bang* (New York: Plenum Press, 1993), 49–67.

19. Smoot and Davidson, 73.

20. Alpher and Herman, 27–30.

21. Parker, 11; Smoot and Davidson, 162.

22. John K. Davies, *Satellite Astronomy* (Chichester, West Sussex, Eng.: Ellis Horwood, 1988), 57, 121.

23. Smoot and Davidson, 89–90, 214–218, 220–223, 298–315.

24. The map of the sky from the COBE data can be seen as a plate in Smoot and Davidson, between pages 150 and 151. Its implications are discussed in Faye Flam, "COBE Sows Cosmological Confusion," *Science,* 3 July 1992, 28–30.

25. Edwin R. Jones and Richard L. Childers, *Contemporary College Physics* (Reading, Mass.: Addison-Wesley, 1993), 674–676; Eric J. Chaisson, *The Hubble Wars* (New York: HarperCollins, 1994), 29–30.

26. John Travis, "Case Closed for a Giant Black Hole," *Science,* 3 June 1994, 1,405; R. Cowen, "Repaired Hubble Finds Giant Black Hole," *Science News,* 4 June 1994, 356–357; Graham P. Collins, "Repaired Hubble Sees Strong Evidence of a Supermassive Black Hole in M87," *Physics Today,* August 1994, 17–20.

27. Parker, 70, 92; John Noble Wilford, "Finding on Universe's Age Poses New Cosmic Puzzle," *New York Times,* 27 October 1994, A1, A14; John Travis, "Hubble War Moves to High Ground," *Science,* 28 October 1994, 539–541.

28. Smoot and Davidson, 115, 158; Paul Davies, *The Last Three Minutes* (New York: Basic Books, 1994), 141–155.

29. Chaisson, 284–286; Bartusiak, 215–216. At the January 1996 meeting of the American Astronomical Society, researchers announced that dark matter consists partly of dead stars, based on observations of gravitational lens effects. See George Johnson, "Dark Matter Lights the Void," *New York Times,* 21 January 1996, section 4, "Week in Review," 1, 4; Sharon Begley, "A Heavenly Host," *Newsweek,* 29 January 1996, 52–53.

30. Smoot and Davidson, 161–163.

Bibliography

Ackerman, Diane. *A Natural History of the Senses.* New York: Vintage, 1990.

Alper, Joseph. "Transillumination: Looking Right Through You." *Science,* 30 July 1993, 560.

Alpher, Ralph, and Robert Herman. "Reflections on Early Work on 'Big Bang' Cosmology." *Physics Today,* August 1988, 24–34.

Angier, Natalie. "New Explanation Given for van Gogh's Agonies." *New York Times,* 21 December 1991, L11.

Applebome, Peter. "Sodden Midwest Is Bracing for More Rain and Floods." *New York Times,* 22 July 1993, A8.

Armitage, E. Liddall. *Stained Glass: History, Technology, and Practice.* Newton, Mass.: Charles T. Branford, 1959.

Arnold, Wilfred Niels. *Vincent van Gogh: Chemicals, Crises, and Creativity.* Boston: Birkhäuser, 1992.

Baeyer, Hans Christian von. *Taming the Atom.* New York: Random House, 1992.

Baltrušaitis, Jurgis. *Anamorphic Art.* New York: Harry N. Abrams, 1977.

Barinaga, Marcia. "Unraveling the Dark Paradox of 'Blindsight.' " *Science,* 27 November 1992, 1,438–1,439.

Barr, E. Scott. "The Infrared Pioneers." *Infrared Physics,* vol. 1 (1961), 1–4.

Bartusiak, Marcia. *Through a Universe Darkly.* New York: HarperCollins, 1993.

Baxandall, Michael. *Patterns of Intention: On the Historical Explanation of Pictures.* New Haven, Conn.: Yale University Press, 1985.

Beerblock, Maurice, Louis Roëlandt, and Georges Charensol. *Vincent van Gogh: Correspondance Générale.* Paris: Gallimard, 1990.

Begley, Sharon. "A Heavenly Host." *Newsweek,* 29 January 1996, 52–53.

Benaron, David A., and David K. Stevenson. "Optical Time-of-Flight and Absorbance Imagining of Biologic Media." *Science,* 5 March 1993, 1,463–1,466.

Berendzen, Richard, Richard Hart, and Daniel Seeley. *Man Discovers the Galaxies.* New York: Science History Publications, 1976.

Bester, Alfred. *The Stars My Destination.* Boston: Gregg Press, 1975.

Bobrick, Benson. *Labyrinths of Iron.* New York: Newsweek Books, 1981.

Bowen, Mary Elizabeth, and Joseph A. Mazzeo. *Writing About Science.* New York: Oxford University Press, 1979.

Boynton, Holmes, editor. *The Beginnings of Modern Science: Scientific Writings of the 16th, 17th, and 18th Centuries.* Roslyn, N.Y.: Walter J. Black, 1948.

Braque, Georges. *Cahier de Georges Braque, 1917–1947.* Paris: Maeght, 1948.

Brenner, David, and Ann Prescott. "Painting with Light." *New Scientist,* vol. 102 (1984), 38–42.

Broad, William J. "Vast Laser Would Advance Fusion and Retain Bomb Experts." *New York Times,* 21 June 1994, B7, B10.

Broad, William J. "From Fantasy to Fact: Space-Based Laser Nearly Ready to Fly." *New York Times,* 6 December 1994, B5–B6.

Bromberg, Joan Lisa. *The Laser in America, 1950–1970.* Cambridge, Mass.: MIT Press, 1991.

Brookhaven Highlights. 1990. Brookhaven National Laboratory, Report No. BNL 52274.

Brookhaven Highlights. 1991. Brookhaven National Laboratory, Report No. BNL 52310.

Brookhaven Highlights. 1992. Brookhaven National Laboratory, Report No. BNL 52360.

Brown, George S. "Imaging the Heart Using Synchrotron Radiation." *Beamline,* vol. 23, no. 3, 1993, 22–28.

Buswell, Guy Thomas. *How People Look at Pictures; A Study of the Psychology of Perception in Art.* Chicago: University of Chicago Press, 1935.

Cajori, Florian. *A History of Physics.* New York: Dover Publications, 1962.

Carey, John, and Neil Gross. "The Light Fantastic." *Business Week,* 10 May 1993, 44–50.

Cerf, Christopher, and Victor Navasky. *The Experts Speak.* New York: Pantheon, 1984.

Chaisson, Eric J. *The Hubble Wars.* New York: HarperCollins, 1994.

Christianson, Gale E. *In the Presence of the Creator: Isaac Newton and His Times.* New York: The Free Press, 1984.

Collins, Graham P. "Repaired Hubble Sees Strong Evidence of a Supermassive Black Hole in M87." *Physics Today,* August 1994, 17–20.

Cotterill, Rodney. *The Cambridge Guide to the Material World.* Cambridge: Cambridge University Press, 1989.

Cowen, R. "Repaired Hubble Finds Giant Black Hole." *Science News,* 4 June 1994, 356–357.

Craxton, R. Stephen, Robert L. McCrory, and John M. Soures. "Progress in Laser Fusion." *Scientific American,* August 1986, 68–79.

Crick, Francis, and Christof Koch. "The Problem of Consciousness." *Scientific American,* September 1992, 153–159.

D'Harnoncourt, Anne, and Walter Hopps. *Etant Donnés: 1° La Chute d'eau; 2° Le gaz d'éclairage: Reflections on a New Work by Marcel Duchamp.* Philadelphia: Philadelphia Museum of Art, 1973.

Davies, John K. *Satellite Astronomy.* Chichester, West Sussex, Eng.: Ellis Horwood, 1988.

Davies, Paul. *The Last Three Minutes.* New York: Basic Books, 1994.

De Camp, L. Sprague. *The Ancient Engineers.* New York: Ballantine, 1962.

————. *Heroes of American Invention.* New York: Barnes and Noble, 1993.

Dennett, Daniel C. *Consciousness Explained.* Boston: Little, Brown, 1991.

Deregowski, Jan B. "Pictorial Perception and Culture." *Scientific American,* November 1972, 82–88.

Deutsch, David, and Michael Lockwood. "The Quantum Physics of Time Travel." *Scientific American,* March 1994, 68–74.

Donnelly, Marian Card. *A Short History of Observatories.* Eugene: University of Oregon, 1973.

Dowling, John E. *The Retina: An Approachable Part of the Brain.* Cambridge, Mass.: Belknap Press, 1987.

Drake, Stillman. *Discoveries and Opinions of Galileo.* Garden City, N.Y.: Doubleday, 1957.

Durant, Frederick C., and Ron Miller. *Worlds Beyond: The Art of Chesley Bonestell.* Norfolk, Va.: Donning, 1983.

Edwards, Terry. *Fiber-Optic Systems: Network Applications.* Chichester, West Sussex, Eng.: John Wiley and Sons, 1989.

Einstein, Albert. *Relativity: The Special and the General Theory.* New York: Wings Books, 1961.

————. *Autobiographical Notes.* P. A. Schilpp, translator and editor. La Salle, Ill.: Open Court Publishing, 1979.

Escher, M. C., *The Graphic Work of M. C. Escher.* New York: Ballantine Books, 1960.

Essick, Robert N., editor. *The Visionary Hand: Essays for the Study of William Blake's Art and Aesthetics.* Los Angeles: Hennessey and Ingalls, 1973.

Essick, Robert N. *William Blake, Printmaker.* Princeton, N.J.: Princeton University Press, 1980.

Fahie, J. J. *Galileo: His Life and Work.* London: John Murray, 1903.

Feller, Robert L. *Artists' Pigments: A Handbook of Their History and Characteristics.* Washington, D.C.: National Gallery of Art, 1986.

Feynman, Richard P. *QED: The Strange Theory of Light and Matter.* Princeton, N.J.: Princeton University Press, 1985.

Feynman, Richard P., Robert B. Leighton, and Matthew Sands. *The Feynman Lectures on Physics, Vols. I, II, and III.* Reading, Mass.: Addison-Wesley, 1966.

Fiberoptics Components Handbook. *Laser Focus World,* June 1994, S5–S40.

Flam, Faye. "COBE Sows Cosmological Confusion." *Science,* 3 July 1992, 28–30.

Freedman, David H. "Theorists to the Quantum Mechanical Wave: 'Get Real.' " *Science,* 12 March 1993, 1,542–1,543.

Freeman, Jude. *The Fauve Landscape.* Los Angeles: Los Angeles County Museum of Art, 1990.

Friedel, Robert, and Paul Israel. *Edison's Electric Light.* New Brunswick, N.J.: Rutgers University Press, 1986.

Furness, S. M. M. *Georges de la Tour of Lorraine.* London: Routledge and Kegan Paul, 1949.

Gibbs, W. Wayt. "Light Motif." *Scientific American,* April 1993, 116–117.

Gjersten, Derek. *The Newton Handbook.* London: Routledge and Kegan Paul, 1986.

Glanz, James. "DOE Lifts Veil of Secrecy from Laser Fusion." *Science,* 17 December 1993, 1,811–1,812.

Gombrich, E. H., Julian Hochberg, and Max Black. *Art, Perception, and Reality.* Baltimore: Johns Hopkins Press, 1972.

Goodrich, Lloyd. *Edward Hopper.* New York: Abradale Press, 1993.

Gosling, Nigel. *Gustave Doré.* New York: Praeger, 1973.

Grady, Denis. "The Vision Thing: Mainly in the Brain." *Discover,* June 1993, 57–66.

Grangier, P., G. Roger, and A. Aspect. "Experimental Evidence for a Photon Anticorrelation Effect on a Beam Splitter: A New Light on Single-Photon Interferences." *Europhysics Letters,* vol. 1 (1986), 173–179.

Gregory, Richard R. *Eye and Brain: The Psychology of Seeing.* Princeton, N.J.: Princeton University Press, 1990.

Hafele, J. C., and Richard E. Keaton. "Around the World Atomic Clocks: Predicted Relativistic Time Gains." *Science,* vol. 177 (1972), 166–168.

―――. "Around the World Atomic Clocks: Observed Relativistic Time Gains." *Science,* vol. 177 (1972), 168–170.

Haglund, Michael M., George A. Ojemann, and Daryl W. Hochman. "Optical Imaging of Epileptoform and Functional Activity in Human Cerebral Cortex." *Nature,* vol. 358 (1992), 668–671.

Hecht, Eugene, and Alfred Zajac. *Optics.* Reading, Mass.: Addison-Wesley, 1979.

Hecht, Jeff. "Military Lasers: The Incredible Becomes Credible." *Laser Focus World,* July 1994, 57–58, 60.

Hellmuth, T., H. Walther, A. Zajonc, and W. Schleich. "Delayed-Choice Experiments in Quantum Interference." *Physical Review A,* vol. 35 (1987), 2,532–2,541.

Henderson, Linda Dalrymple. *The Fourth Dimension and Non-Euclidean Geometry in Modern Art.* Princeton, N.J.: Princeton University Press, 1983.

Hobbs, Robert. *Edward Hopper.* New York: Harry N. Abrams, 1987.

Holton, W. Conrad. "The Light Brigade." *Photonics Spectra,* June 1994, 76–81.

Home, Dipankar, and John Gribbin. "What Is Light?" *New Scientist,* 2 November 1991, 30–33.

Hopper, Edward. *Edward Hopper: Forty Masterworks.* New York: W. W. Norton, 1988.

Horgan, John. "Quantum Philosophy." *Scientific American,* July 1992, 94–101.

Hubel, David H. *Eye, Brain, and Vision.* New York: Scientific American Library, 1988.

Hulbert, S. L., and N. M. Lazarz, editors. *National Synchrotron Light Source Annual Report 1991,* vols. 1, 2. Brookhaven National Laboratory, Report No. BNL 52317/UC400, April 1992.

Irwin, Robert, and Russell Ferguson. *Robert Irwin.* Los Angeles: Museum of Contemporary Art, 1993.

Jaffe, Bernard. *Michelson and the Speed of Light.* Garden City, N.Y.: Anchor Books, 1960.

Jirat-Wasiutynski, Vojtech, H. Travers Newton, Eugene Farrell, and Richard Newman. *Vincent van Gogh's "Self-portrait Dedicated to Paul Gauguin": An Historical and Technical Study.* Cambridge, Mass.: Center for Conservation and Technical Studies, Harvard University Art Museums, 1984.

Johnson, George. "Dark Matter Lights the Void." *New York Times,* 21 January 1996, section 4, "Week in Review," 1, 4.

Jones, Edwin R., and Richard L. Childers. *Contemporary College Physics.* Reading, Mass.: Addison-Wesley, 1993.

Kemp, Martin. *The Science of Art.* New Haven: Yale University Press, 1990.

Kendall, Richard. "Degas and the Contingency of Vision." *The Burlington Magazine,* March 1988, 180–197.

Khosla, Rajinder P. "From Photons to Bits." *Physics Today,* December 1992, 42–49.

Kiernan, Vincent. "Military Lasers Face Cloudy Future." *Laser Focus World,* May 1994, 71–72.

Kimmelman, Michael. "A Face in the Gallery of Picasso's Muses Is Given a New Name." *New York Times,* 21 April 1994, B1, B2.

King, Henry C. *The History of the Telescope.* London: Charles Griffin, 1955.

Kitson, Michael. *The Complete Paintings of Caravaggio.* New York: Harry N. Abrams, 1967.

Kowler, Eileen, editor. *Eye Movements and Their Role in Visual and Cognitive Processes.* Amsterdam: Elsevier, 1990.

Kuthy, Sandor, and Kuniko Satonubu. *Sonia & Robert Delaunay.* Stuttgart: Verlag Gerd Hatje, 1991.

Lane, William Cooledge, and Nina E. Browne, editors. *American Library Association Portrait Index.* Washington, D.C.: Government Printing Office, 1906.

Learner, Richard. *Astronomy Through the Telescope.* New York: Van Nostrand Reinhold, 1981.

Leary, Warren E. "Optical Imaging Offers Gentler Way to Monitor Human Brain at Work." *New York Times,* 25 August 1992, B6.

Leeman, Fred, Joost Elffers, and Michael Schuyt. *Hidden Images: Games of Perception, Anamorphic Art, Illusion from the Renaissance to the Present.* New York: Harry N. Abrams, 1976.

Leutwyler, Kristin. "Optical Tomography." *Scientific American,* January 1994, 147–148.

Ley, Willy. *Watchers of the Skies.* New York: Viking Press, 1963.

Ley, Willy, and Chesley Bonestell. *The Conquest of Space.* New York: Viking Press, 1950.

Livingston, Dorothy Michelson. *The Master of Light: A Biography of Albert A. Michelson.* New York: Charles Scribner's Sons, 1973.

Loudon, Rodney. *The Quantum Theory of Light.* Oxford: Clarendon Press, 1981.

Lurie, Alison. *The Language of Clothes.* New York: Vintage Books, 1983.

Marvin, Carolyn. *When Old Technologies Were New.* New York: Oxford University Press, 1988.

Mason, Peter. *The Light Fantastic.* Ringwood, Victoria, Australia: Penguin Australia, 1981.

Meehan, Beth Ann. "Seeing Red: It's Written in Your Genes." *Discover,* June 1993, 66.

National Research Council. *Photonics: Maintaining Competitiveness in the Information Era.* Washington, D.C.: National Academy Press, 1988.

Newton, Isaac. *Opticks.* New York: Dover Publications, 1952.

"Nova Nears Completion." *Physics Today,* September 1984, 20.

O'Dea, William T. *The Social History of Lighting.* London: Routledge and Kegan Paul, 1958.

Ovid and Rolfe Humphries, translator. *Metamorphoses.* Bloomington: Indiana University Press, 1955.

Palter, Robert, editor. *The Annus Mirabilis of Sir Isaac Newton, 1666–1966.* Cambridge, Mass.: M.I.T. Press, 1970.

Parker, Barry. *The Vindication of the Big Bang.* New York: Plenum Press, 1993.

Passingham, W. J. *Romance of London's Underground.* New York: Benjamin Blom, 1972.

Paterson, Harriet. "Raphael, the Duke and £20 Million." *The Sunday Telegraph,* 9 February 1992, 115.

Patton, Phil. "The Pixels and Perils of Getting Art on Line." *New York Times,* 7 August 1994, Arts and Leisure 1, 31.

Pennisi, E. "Synchrotron Beam Sees Record Tiny Crystal." *Science News,* 14 September 1991, 164.

Perkowitz, Sidney. "The War Science Waged." *Washington Post,* 3 March 1991, C2.

Peterson, Ivars. "Chaos in Spacetime." *Science News,* vol. 144 (1993), 376–377.

Preses, Jack M., J. Robb Grover, Åke Kvick, and Michael G. White. "Chemistry with Synchrotron Radiation." *American Scientist,* September–October 1990, 424–437.

Ramón y Cajal, Santiago. *Recollections of My Life.* E. Horne Craigie, translator. Philadelphia: The American Philosophical Society, 1937.

Renonciat, Annie. *La Vie et l'Oeuvre de Gustave Doré.* Paris: ACR Edition, 1983.

Robertson, Alan R. "Color Perception." *Physics Today,* December 1992, 24–29.

Rock, Irvin. *Perception.* New York: Scientific American Library, 1984.

Ronchi, Vasco. *The Nature of Light.* V. Barocas, translator. Cambridge, Mass.: Harvard University Press, 1970.

Rossotti, Hazel. *Colour: Why the World Isn't Grey.* Princeton, N.J.: Princeton University Press, 1983.

Rubin, William. "The Pipes of Pan: Picasso's Aborted Love Song to Sara Murphy." *ARTNews,* May 1994, 138–147.

Ruspoli, Mario. *The Cave of Lascaux.* New York: Harry N. Abrams, 1987.

Sacks, Oliver. "To See or Not to See." *The New Yorker,* 10 May 1993, 59–73.

Saleh, Bahaa E. A., and Malvin Carl Teich. *Fundamentals of Photonics.* New York: John Wiley and Sons, 1991.

Schama, Simon. *Citizens: A Chronicle of the French Revolution.* New York: Alfred A. Knopf, 1989.

Schapiro, Meyer. *Van Gogh.* New York: Harry N. Abrams, 1983.

Schewe, Philip P., and Ben P. Stein. *American Physical Society News: Physics News in 1993,* May 1994.

Schilpp, Paul Arthur, editor. *Albert Einstein: Philosopher-Scientist.* Evanston, Ill.: The Library of Living Philosophers, 1949.

Schivelbusch, Wolfgang. *Disenchanted Night: The Industrialization of Light in the Nineteenth Century.* Berkeley and Los Angeles: University of California Press, 1988.

Schnapf, Julie I., and Denis A. Baylor. "How Photoreceptor Cells Respond to Light." *Scientific American,* April 1987, 40–47.

Sharov, Alexander S., and Igor D. Navikov. *Edwin Hubble: The Discoverer of the Big Bang Universe,* Vitaly Kisin, translator. Cambridge: Cambridge University Press, 1993.

Shepp, James W., and Daniel B. Shepp. *Shepp's World's Fair Photographed.* Chicago: Globe Bible Publishing, 1893.

Shimazu, Michael. "Optical Computing Coming of Age." *Photonics Spectra,* November 1992, 66–74.

Shlain, Leonard. *Art and Physics: Parallel Visions in Space, Time, and Light.* New York: William Morrow, 1991.

Sinnott, Roger. "Astronomical Computing: Van Gogh, Two Planets, and the Moon." *Sky and Telescope,* October 1988, 406–408.

Smith, Bradford L. "World's Largest Mirror to Paris." *Photonics Spectra,* September 1993, 30–31.

Smith, R. A., F. E. Jones, and R. P. Chasmar. *The Detection and Measurement of Infrared Radiation.* Oxford: The Clarendon Press, 1968.

Smoot, George, and Keay Davidson. *Wrinkles in Time.* New York: William Morrow, 1993.

Speer, Albert. *Albert Speer: Architektur: Arbeiten 1933–1942.* Frankfurt am Main: Propyläen, 1978.

———. *Albert Speer: Architecture, 1932–1942.* Brussels: Archives d'Architecture Moderne, 1985.

———. *Inside the Third Reich.* New York: Macmillan, 1970.

Stoerig, Petra. "Sources of Blindsight." *Science,* 23 July 1993, 493–495.

Stone, Irving, editor. *Dear Theo.* Boston: Houghton Mifflin Company, 1937.

Stryer, Lubert. "The Molecules of Visual Excitation." *Scientific American,* July 1987, 42–51.

Szarkowski, John. *The Photographer's Eye.* New York: Museum of Modern Art/Doubleday, 1966.

Taubes, Gary. "Laser Fusion Catches Fire." *Science,* 3 December 1993, 1,504–1,506.

Taylor, G. I. "Interference Fringes with Feeble Light." *Proceedings of the Cambridge Philosophical Society,* vol. 15 (1909), 114–115.

"Thermal Imaging Joins Lyme-Disease Battle." *Photonics Spectra,* July 1994, 20, 22.

Thuillier, Jacques. *Georges de La Tour.* Paris: Flammarion, 1992.

Thwing, Leroy. *Flickering Flames.* Rutland, Vt.: Charles E. Tuttle, 1958.

Tipler, Paul A. *Physics for Scientists and Engineers.* New York: Worth, 1991.

Travis, John. "Laser Replication of Rare Art." *Science,* 15 May 1992, 969.

———. "Case Closed for a Giant Black Hole." *Science,* 3 June 1994, 1,405.

———. "Hubble War Moves to High Ground." *Science,* 28 October 1994, 539–541.

Troy, Charles T. "Laser Facility Will Seek to Duplicate the Big Bang." *Photonics Spectra,* May 1994, 31.

Turnbull, H. W., editor. *The Correspondence of Isaac Newton, Volume 1, 1661–1675.* Cambridge: Cambridge University Press, 1959.

Turrell, James, Barbara Haskell, and Melinda Wortz. *Light and Space.* New York: Whitney Museum of American Art, 1980.

Turrell, James, Julia Brown, and Craig E. Adcock. *Occluded Front, James Turrell.* Los Angeles: Fellows of Contemporary Art, 1985.

Vigier, J. P. "From Descartes and Newton to Einstein and de Broglie." *Foundations of Physics,* vol. 23 (1993), 1–4.

Wade, Nicholas J., and Michael Swanston. *Visual Perception: An Introduction.* London: Routledge, 1991.

Wagner, Fritz. *Isaac Newton im Zwielicht zwischen Mythos und Forschung.* Freiburg: Alber, 1976.

Waldrop, M. Mitchell. "Computing at the Speed of Light." *Science,* 22 January 1993, 456.

Wallace, Doris B., and Howard E. Gruber. *Creative People at Work.* New York: Oxford University Press, 1989.

Walther, Ingo F., and Rainer Metzger. *Vincent van Gogh: The Complete Paintings.* Cologne Benedikt Tascher, 1990.

Warner, Deborah Jean. *Alvan Clark and Sons: Artists in Optics.* Washington, D.C.: Smithsonian Institution, 1968.

Weaver, Warren, editor. *The Scientists Speak.* New York: Boni and Gaer, 1947.

Weinberg, Steven. *The First Three Minutes.* New York: Basic Books, 1988.

Wertenbaker, Lael. *The Eye: Window to the World.* New York: Torstar Books, 1984.

Westfall, Richard S. *The Life of Isaac Newton.* Cambridge: Cambridge University Press, 1993.

Wheaton, Bruce R. *The Tiger and the Shark: Empirical Roots of Wave-Particle Dualism.* Cambridge: Cambridge University Press, 1983.

Whitney, Charles A. "The Skies of Vincent van Gogh." *Art History,* September 1986, 351–362.

Wilford, John Noble. "Finding on Universe's Age Poses New Cosmic Puzzle." *New York Times,* 27 October 1994, A1, A14.

Wilford, John Noble. "New Galaxy May Shed Light on Universe." *New York Times,* 1 February 1996, A12.

Williams, G. P., R. Budhani, C. J. Hirschmugl, G. L. Carr, S. Perkowitz, B. Lou, and T. R. Yang. "Infrared Synchrotron Radiation Transmission Spectroscopy of YBaCuO in the Gap and Supercurrent Region." *Physical Review B,* vol. 41 (1990), 4,752–4,755.

Williams, Gwyn. *Using the Light Fantastic.* Brookhaven National Laboratory, Report No. BNL-51988/UC-28, Number 211, January 16, 1985.

Winick, Herman, and Gwyn Williams. "Overview of Synchrotron Radiation Sources World-Wide." *Synchrotron Radiation News,* vol. 4, no. 5 (1991), 23–26.

Winick, Herman. "Synchrotron Radiation." *Scientific American,* November 1987, 88–99.

Wolf, Emil. "Einstein's Researches on the Nature of Light," *Optics News,* vol. 5 (1979), 24–39.

Wright, Christopher. *Georges de La Tour.* Oxford: Phaidon, 1977.

Yen, W. M., and M. D. Levenson, editors. *Lasers, Spectroscopy, and New Ideas.* Berlin: Springer-Verlag, 1987.

Zajonc, Arthur. *Catching the Light.* New York: Bantam Books, 1993.

Zava Boccazi, Franca. *Pittoni: L'Opera Completa.* Venice: Alfieri, 1979.

Zeki, Semir. "The Visual Image in Mind and Brain." *Scientific American,* September 1992, 69–76.

Zurcher, Bernard. *Vincent van Gogh: Art, Life, and Letters.* New York: Rizzoli, 1985.

Art Credits

Georges de la Tour, *The Penitent Magdalen.* The Metropolitan Museum of Art, gift of Mr. and Mrs. Charles Wrightsman, 1978. (1978.517) Photograph copyright 1997 The Metropolitan Museum of Art.

Chapter Six
Vincent van Gogh, *The Langlois Bridge at Arles.* Wallraf-Richartz-Museum, Cologne. Photograph, Rheinisches Bildarchiv.

Vincent van Gogh, *Self-Portrait Dedicated to Paul Gauguin.* Courtesy of the Fogg Art Museum, Harvard University Art Museum, bequest from the collection of Maurice Wertheim, Class of 1906. Photograph, David Mathews. Copyright President and Fellows of Harvard College, Harvard University.

M.C. Escher, *Hand Holding Reflecting Sphere.* Copyright 1998 Cordon Art B.V. - Baarn - Holland. All rights reserved.

Chapter Seven
Gwyn Williams, National Synchrotron Light Source. Brookhaven National Laboratory, Upton, New York.

Chapter Eight
Vincent van Gogh, *The Starry Night.* The Museum of Modern Art, New York. Acquired through the Lillie P. Bliss Bequest. Photograph copyright 1998 The Museum of Modern Art, New York.

National Aeronautics and Space Administration, Gaseous Pillars in M16— Eagle Nebula, Hubble Space Telescope photograph.

Index

Gravitational forces, 10, 76-78, 180, 190-91
Great Wall, 187
Great Wave off Kanagawa, The (Hokusai), 23
Greeks, ancient, 46-47
Grimaldi, Francesco Mario, 52, 69
Guiding waves, 88

Hand Holding Reflecting Sphere (Escher), 123
Haskell, Francis, 56
Hawking, Stephen, 53
Heisenberg, Werner, 12, 87
Heliograph, 163
Helium formation, 8
Helmholtz, Hermann von, 18, 93, 102
Henderson, Linda Dalrymple, 70
Henderson, Sir Neville, 111
Heredity, 27-28
Herman, Robert, 185-86
Herschel, John, 130
Herschel, William, 14, 39, 171, 174-75, 177
Hertz, Heinrich, 20, 39, 63
Hertz units, 20
Heuring, Vincent, 150
Hobbs, Robert, 106
Hokusai, 23
Holbein, Hans, 123
Hooke, Robert, 52, 55, 131
Hopper, Edward, 4, 16-17, 106
Hoyle, Fred, 185
Huang, Alan, 142, 150
Hubble, Edwin Powell, 14, 168, 181, 182-85
Hubble Space Telescope, 189-90
Hubel, David, 19, 31
Huggins, William, 178-79, 181
Humason, Milton, 184
Huygens, Christian, 50, 58
Hydrogen formation, 8

Imbusch, G. F., 117

Immune system, 40
Incandescent electric light, 99-103
Induction, 60, 61
Information-carrying capacity, 145-46
Infrared light, 5, 9, 14, 40, 41, 133, 149, 164-65, 177
Interference principle
    photons and, 83-84
    wave theory and, 52, 58, 66-67
Interferometer, 66-67, 68
International Ultraviolet Explorer (IUE), 188
Internet, 13, 144, 148
Invisible light, 5, 39-42
Irwin, Robert, 13, 109

Jablochkov, Paul, 99
James, William, 33
Jannsen, Zacharias, 131
Javal, L. E., 22
Jefferson, Thomas, 97
Jordan, Harry, 150

Kant, Immanuel, 18, 175, 181
Karloff, Boris, 150
Kelvin, Lord, 68
Kendall, Richard, 38
Kepler, Johannes, 118, 128, 170
Killers, The (film), 3
Klee, Paul, 4
Kneller, Sir Godrey, 56
Koch, Christof, 33
Kubrick, Stanley, 44

Lamps, 94-99
Lancaster, Burt, 3
Lang, Fritz, 150
Langlois Bridge at Arles, The (van Gogh), 121, 123
Language of Clothes, The (Lurie), 36
Lartigue, Jacques Henri, 73
Lascaux cave paintings, 16, 36
Lasers
    fusion experiments with, 151-52